Machine Learning: Foundations, Methodologies, and Applications

Books published in this series focus on the theory and computational foundations, advanced methodologies and practical applications of machine learning, ideally combining mathematically rigorous treatments of a contemporary topics in machine learning with specific illustrations in relevant algorithm designs and demonstrations in real-world applications. The intended readership includes research students and researchers in computer science, computer engineering, electrical engineering, data science, and related areas seeking a convenient medium to track the progresses made in the foundations, methodologies, and applications of machine learning.

Topics considered include all areas of machine learning, including but not limited to:

- Decision tree
- Artificial neural networks
- Kernel learning
- Bayesian learning
- Ensemble methods
- Dimension reduction and metric learning
- Reinforcement learning
- Meta learning and learning to learn
- Imitation learning
- Computational learning theory
- Probabilistic graphical models
- Transfer learning
- Multi-view and multi-task learning
- Agents and Multi-Agent Systems
- Graph neural networks
- Generative adversarial networks
- Federated learning
- Large Language Models
- Multimodal Learning
- Transformer/Diffusion Models
- Generative Artificial Intelligence
- Bio-inspired Learning Models
- Embodied AI
- Explainable AI and Ethics

This series includes monographs, introductory and advanced textbooks, and state-of-the-art collections. Furthermore, it supports Open Access publication mode.

Yang Yu · Hong Qian · Yi-Qi Hu

Derivative-Free Optimization

Theoretical Foundations, Algorithms, and Applications

Springer

Yang Yu
Nanjing University
Nanjing, China

Hong Qian
East China Normal University
Shanghai, China

Yi-Qi Hu
Huawei Technologies Co., Ltd.
Nanjing, China

ISSN 2730-9908 ISSN 2730-9916 (electronic)
Machine Learning: Foundations, Methodologies, and Applications
ISBN 978-981-96-5928-9 ISBN 978-981-96-5929-6 (eBook)
https://doi.org/10.1007/978-981-96-5929-6

This Springer imprint is published by the registered company Springer Nature Singapore Pte Ltd.
The registered company address is: 152 Beach Road, #21-01/04 Gateway East, Singapore 189721, Singapore

Preface

The pursuit of solving optimization problems has long been a cornerstone in the realms of computer science and artificial intelligence. While specialized algorithms tailored to specific problems have been developed, there has also been a significant emergence of general-purpose optimization methods. These methods, including random search, simulated annealing, and evolutionary algorithms, are often crafted based on heuristic principles. Consequently, their properties are predominantly explored through empirical studies, with a comprehensive theoretical understanding still largely out of reach.

Yang Yu, the first author of this monograph, previously co-authored a book entitled *Evolutionary Learning: Advances in Theories and Algorithms* with Prof. Zhi-Hua Zhou and Chao Qian, which was published by Springer in 2019. In that work, the central theoretical focus was on running time analysis, which evaluates the time and sample complexity required to find an optimal solution. Subsequently, Yang developed an interest in exploring alternative theoretical foundations for general-purpose optimization methods, complementing the well-established theories of machine learning. Given that machine learning is underpinned by a solid statistical framework, it naturally raises the question: can we establish a similarly strong theoretical foundation for optimization methods from a statistical perspective? Specifically, such a theory could elucidate how these optimization methods approximate optimal solutions.

Collaborating with the second author, Hong Qian, who was Yang's Ph.D. student in 2013, a preliminary framework was developed. Initially, this framework was designed to meet certain theoretical desiderata. Surprisingly, the resulting algorithm also demonstrated competitive practical performance against some state-of-the-art methods. The third author, Yi-Qi Hu, who became Yang's Ph.D. student in 2015, joined the effort to further refine and extend the framework and enhance the algorithm. This monograph provides a comprehensive overview of the authors' research, encompassing the development of the framework, algorithm design, and practical applications.

The monograph is structured into four parts. Part I briefly introduces derivative-free optimization within the context of machine learning. Part II presents the classification-based optimization framework along with its basic algorithm. To tackle practical challenges such as sequential execution, high-dimensionality, noisy evaluations, and large-scale parallel execution, Part III introduces several variants of the basic algorithm. This part also includes an introduction to ZOOpt, a general optimization toolbox built on classification-based optimization algorithms. Part IV showcases various applications of classification-based optimization in the field of automatic machine learning, including hyper-parameter selection, algorithm selection, and neural architecture search.

The authors extend their heartfelt gratitude to their families, friends, and collaborators for their unwavering support and contributions.

Nanjing, China — Yang Yu
Shanghai, China — Hong Qian
Nanjing, China — Yi-Qi Hu
December 2024

Competing Interests The authors have no competing interests to declare that are relevant to the content of this manuscript.

Contents

Part I Introduction

1 **Introduction** ... 3
1.1 Machine Learning ... 4
1.2 Derivative-Free Optimization (DFO) ... 5
1.2.1 Structure of DFO Algorithms ... 5
1.2.2 Development of DFO Algorithms ... 6
1.3 Automatic Machine Learning ... 7
1.4 Organization of the Book ... 8
References ... 8

2 **Preliminaries** ... 11
2.1 Evolutionary Algorithms ... 11
2.1.1 $(\mu + \lambda)$-EA ... 12
2.1.2 $(\mu/\mu, \lambda)$-ES ... 12
2.2 Estimation of Distribution Algorithms ... 13
2.3 Bayesian Optimization ... 15
2.4 Running Time Analysis ... 16
2.5 No Free Lunch in Optimization ... 17
References ... 18

Part II Classification-Based Derivative-Free Optimization

3 **Framework** ... 23
3.1 Sampling and Learning Framework ... 24
3.2 Casting Previous DFO Methods Into the SAL Framework ... 25
3.2.1 Estimation of Distribution Algorithms ... 25
3.2.2 Bayesian Optimization ... 26
3.2.3 Evolutionary Algorithms ... 26
3.2.4 Other DFO Methods ... 27

3.3 Sampling and Classification Framework 28
3.4 Summary 30
References 31

4 Theoretical Foundation 33
4.1 Problem Setting and Notations 34
4.2 (ϵ, δ)-Query Complexity 35
4.3 Performance Bound for SAL Framework 36
4.4 Performance Bound for SAC Framework 38
4.5 Error-Target Dependence and Shrinking Rate 40
4.6 Functions with Local Lipschitz Continuity 43
4.7 Functions with Bounded Packing and Covering Numbers 46
4.8 Summary 48
References 48

5 Basic Algorithm 49
5.1 The RACOS Optimization Algorithm 50
5.2 Empirical Study on Testing Functions 52
5.3 Empirical Study on Clustering Task 53
5.4 Empirical Study on Classification with Ramp Loss 55
5.5 Summary 56
References 57

Part III Practical Extensions

6 Optimization in Sequential Mode 61
6.1 Sequential Classification Model Based Algorithm 62
6.2 Theoretical Analysis 63
6.3 Empirical Study 65
6.3.1 Optimization on Synthetic Functions 65
6.3.2 Direct Policy Search on Reinforcement Learning Tasks 68
6.4 Summary 71
References 71

7 Optimization in High-Dimensional Search Space 73
7.1 Functions with Low Effective Dimension 74
7.1.1 Random Embedding for Low Effective Dimension Problems 74
7.2 Optimal ε-Effective Dimension 75
7.2.1 Random Embedding for Problems with Low Optimal ε-Effective Dimension 76
7.2.2 Optimization with Random Embedding 76
7.3 Sequential Random Embeddings 77
7.3.1 Less Greedy SRE 79

7.4 Empirical Study ... 80
7.4.1 Experimental Setup ... 80
7.4.2 Synthetic Functions ... 80
7.4.3 Classification with Ramp Loss ... 83
7.5 Summary ... 85
References ... 85

8 Optimization Under Noise ... 87
8.1 Value Suppression ... 88
8.2 The SSRacos Algorithm ... 89
8.3 Empirical Study ... 91
8.3.1 Synthetic Functions ... 91
8.3.2 Controlling Tasks in OpenAI Gym ... 94
8.3.3 Hyper-Parameter Analysis ... 97
8.4 Summary ... 97
References ... 99

9 Optimization with Parallel Computing ... 101
9.1 The Asynchronous SRacos (ASRacos) Algorithm ... 102
9.2 Theoretical Analysis ... 104
9.3 Empirical Study ... 105
9.3.1 On Synthetic Functions ... 106
9.3.2 On Controlling Tasks in OpenAI Gym ... 107
9.4 Summary ... 112
References ... 113

10 Toolbox: ZOOpt ... 115
10.1 Methods in ZOOpt ... 115
10.2 Usage ... 117
10.3 Experiments ... 121
10.3.1 Results on Optimizing Synthetic Functions ... 122
10.3.2 Results on Classification Tasks with Ramploss ... 124
10.3.3 Results on Direct Policy Search for OpenAI Controlling Tasks ... 125
10.4 Summary ... 128
References ... 128

Part IV Application to Automatic Machine Learning

11 Experienced Optimization: Acceleration in Hyper-Parameter Optimization ... 133
11.1 Experienced Optimization for Hyper-Parameter Optimization ... 134
11.2 The ExpSRacos and AdaSRacos Algorithms ... 134
11.2.1 ExpSRacos ... 135
11.2.2 AdaSRacos ... 137

11.3 Empirical Study ... 140
11.3.1 Synthetic Tasks ... 140
11.3.2 Hyper-Parameter Optimization Tasks ... 142
11.4 Summary ... 143
References ... 145

12 Multi-fidelity Optimization: Acceleration in Hyper-Parameter Evaluation ... 147
12.1 Multi-fidelity Optimization for Hyper-Parameter Optimization ... 148
12.2 The TSESRACOS Algorithm ... 149
12.2.1 Multi-fidelity Optimization Framework ... 149
12.2.2 Transfer Series Expansion (TSE) ... 150
12.3 Empirical Study ... 152
12.3.1 Experimental Setup ... 152
12.3.2 Empirical Analysis ... 154
12.4 Summary ... 157
References ... 157

13 Stepwise Optimization: Cascaded Algorithm Selection ... 159
13.1 Stepwise Optimization with Algorithm Selection ... 160
13.2 The ER-UCB Algorithm ... 161
13.2.1 Extreme Region Target and Extreme Region Regret ... 162
13.2.2 ER-UCB on Stationary Distributions ... 163
13.2.3 ER-UCB on Non-stationary Distributions ... 164
13.2.4 Theoretical Results ... 165
13.3 Empirical Study ... 166
13.3.1 Synthetic Tasks ... 166
13.3.2 AutoML Tasks ... 169
13.4 Summary ... 174
References ... 174

14 Calculation Operation Optimization: Competition Neural Architecture Search ... 177
14.1 Calculation Operation Optimization with Neural Architecture Search ... 178
14.1.1 NAS Task Formulation ... 179
14.1.2 Topological Structure Enumeration ... 179
14.1.3 Calculation Operation Optimization ... 180
14.2 The CNAS Algorithm ... 182
14.2.1 Block-Based Search ... 183
14.2.2 Experience Reuse ... 183
14.2.3 Experience-Reused CNAS ... 185

14.3 Empirical Study 187
14.3.1 Image Classification Tasks 187
14.3.2 Image Denoising Tasks 190
14.4 Summary 191
References 192

Notations

$\mathbb{R}$	Real number
$\mathbb{N}$	Integer
$(\cdot)^+$	Positive, $(\cdot)$ can be $\mathbb{R}$ or $\mathbb{N}$
$(\cdot)^{0+}$	Non-negative, $(\cdot)$ can be $\mathbb{R}$ or $\mathbb{N}$
x	Scalar
$\boldsymbol{x}$	Vector
$(\cdot, \cdot, \ldots, \cdot)$	Row vector
$(\cdot; \cdot; \ldots; \cdot)$	Column vector
$\mathbf{0}, \mathbf{1}$	All-0s and all-1s row vectors
$\{0, 1\}^n$	Boolean vector space
$\boldsymbol{X}$	Matrix
$(\cdot)^\top$	Transpose of a vector/matrix
X	Set
$\{\cdot, \cdot, \cdots, \cdot\}$	Set by enumeration
$[n]$	Set $\{1, 2, \ldots, n\}$
$\|\cdot\|$	Cardinality of a set
2^X	Power set of X, which consists of all subsets of X
$X - Y$	Complement of Y in X, which consists of elements in X but not in Y
$\Pr(\cdot), \Pr(\cdot\|\cdot)$	Probability and conditional probability
$\mathcal{D}$	Probability distribution
f	Function
$\mathbb{E}_{\cdot\sim\mathcal{D}}[f(\cdot)], \mathbb{E}_{\cdot\sim\mathcal{D}}[f(\cdot)\|\cdot]$	Expectation and conditional expectation of $f(\cdot)$ under distribution $\mathcal{D}$, simplified as $\mathbb{E}[f(\cdot)]$ and $\mathbb{E}[f(\cdot)\|\cdot]$ when the meaning is clear
$\mathbb{I}(\cdot)$	Indicator function which takes 1 if $\cdot$ is true, and 0 otherwise
$\lfloor\cdot\rfloor, \lceil\cdot\rceil$	Floor and ceiling functions which take the greatest/least integer less/greater than or equal to a real number

Part I
Introduction

Chapter 1
Introduction

Abstract This chapter introduces the fundamental concepts of optimization, particularly in the context of machine learning, and explores the role of derivative-free optimization (DFO) in solving complex computational tasks. Optimization is essential for finding optimal solutions within a solution space, and machine learning often involves formulating such problems to learn generalizable models from data. The chapter highlights the importance of DFO, which does not require gradient information and is suitable for problems with discontinuous or non-differentiable objective functions. It outlines the structure of DFO algorithms, their development, and their application in automatic machine learning (AutoML), where they help automate the selection of algorithms and hyper-parameters. The chapter concludes by presenting the organization of the book, which aims to build theoretical foundations for DFO and design practical algorithms for machine learning tasks.

Optimization is pervasive and fundamental in complex computational tasks. It is also an area where computers can be immensely useful to us. An optimization problem involves searching for an optimal solution within a solution space. The solution space often depends on the specific task and is most commonly represented as a vector space $\mathcal{X}$. The quality of a solution is evaluated using an objective function f. A general optimization problem can be written as

$$\boldsymbol{x}^* = \operatorname*{arg\,min}_{x \in \mathcal{X}} f(\boldsymbol{x}), \tag{1.1}$$

where $\boldsymbol{x}^*$ is the optimal solution.

In this book, we are particularly interested in optimization problems that arise in machine learning tasks. This is not only because machine learning involves a wide variety of optimization problems but also because the optimization methods developed in this book draw inspiration from machine learning concepts.

Y. Yu et al., *Derivative-Free Optimization*, Machine Learning: Foundations, Methodologies, and Applications, https://doi.org/10.1007/978-981-96-5929-6_1

1.1 Machine Learning

Machine learning [12] is a subfield of artificial intelligence that studies how to learn generalizable models from data. In supervised learning, a classical machine learning task, a training dataset typically contains examples of input-output pairs, where an instance of the input is called a feature vector and the corresponding output is called a label. A central goal of supervised learning is to learn a model based on the training dataset. A model is a function that maps from the input space to the output space, which can take the form of a decision tree, linear model, artificial neural network, etc. The model is expected to not only be consistent with the training dataset but, more importantly, to be generalizable, meaning it can accurately predict labels for instances outside of the training dataset. The model learning process can be viewed as a search process in a model space, also known as a hypothesis space, to find the model that best fits the training dataset. To ensure generalizability, constraints are commonly incorporated into the search process. Therefore, many supervised learning tasks can be formulated as

$$h^* = \arg\min_{h \in \mathcal{H}} \sum_{(x,y) \in D} \ell(h(x), y) + G(h), \tag{1.2}$$

where h is a model, $\mathcal{H}$ is the hypothesis space, D is the training dataset containing (x, y) example pairs, ℓ is a loss function measuring the difference between the model output $h(x)$ and the label y, and $G(h)$ is a penalty on the model complexity.

Other branches of machine learning can be formulated in a similar manner. For unsupervised clustering, a general formulation is

$$h^* = \arg\min_{h \in \mathcal{H}} \ell(D_1, D_2, \ldots, D_k | D_i = \{h(x) = i\}) + G(h), \tag{1.3}$$

where ℓ is the evaluation criterion on the clusters partitioned by the model. For reinforcement learning, a general formulation is

$$\pi^* = \arg\max_{\pi \in \Pi} \sum_{s \in S} \rho^{\pi}(s) \sum_{a \in A} \pi(a|s) R(s, a) + G(\pi), \tag{1.4}$$

where π is the policy model, Π is the hypothesis space, ρ is the stationary distribution induced by the policy, and R is the reward function.

The key components of machine learning can be observed from the above formulations, including the definition of the model space, the choice of loss function and complexity penalty, and the optimization procedure. In other words, the representation, evaluation, and optimization, as summarized by Domingos [4]. The choice of representation and evaluation define the optimization problem. Meanwhile, the available optimization methods constrain the design choices for representation and evaluation. We can see that a rich toolbox of optimization methods enables a wide range of design choices, leading to machine learning techniques suitable for various situations and requirements.

Although a large body of research focuses on designing representations and evaluations to simplify the optimization task, this book aims to develop general-purpose optimization tools that can support diverse representations and evaluations, including discontinuous functions with many local optima. To this end, we consider derivative-free optimization.

1.2 Derivative-Free Optimization (DFO)

1.2.1 Structure of DFO Algorithms

To handle a broad range of optimization tasks, we do not assume the objective function f in Eq. (1.1) to be linear, convex, differentiable, or even explicitly known. Instead, we only assume access to the objective function value $f(x)$ for any given solution x, without requiring other information such as the gradient. This requirement characterizes *derivative-free optimization* (DFO), also known as *zeroth-order optimization* or *black-box optimization*.

The simplest DFO method is perhaps random search, which generates solutions uniformly at random and evaluates their objective values. The solution with the best objective value is selected as the final output. Random search has been widely used for hyper-parameter tuning in machine learning algorithms.

Despite its simplicity, random search shares a common framework with other derivative-free optimization methods, as shown in Fig. 1.1. The framework consists of three main components. The sampling component draws solutions from a distribution. In random search, the distribution $\mathcal{T}$ is simply the uniform distribution over the solution space, and solutions are sampled one at a time. The sampled solutions are evaluated to obtain their objective values, and the best solution found so far is recorded as the algorithm's output. Based on the evaluated solutions, a DFO algorithm updates its sampling distribution $\mathcal{T}$ to guide the search towards better solutions in the next iteration. Random search does not update the sampling distribution, whereas many DFO algorithms differ in how they represent and adapt the sampling distribution.

One may wonder why this DFO framework can solve optimization problems. Let's reconsider random search. Despite its simplicity, random search will converge to the global optimum as it explores the entire solution space. In other words, random search is complete in its exploration. However, random search is inefficient and can take a very long time to find the optimal solution. Suppose the desired solutions, i.e., solutions that are sufficiently good, reside in a region $\mathcal{X}^*$. Each time we randomly sample a solution from $\mathcal{X}$, the probability of finding a desired solution is the proportion of $\mathcal{X}^*$ in the solution space, i.e., $|\mathcal{X}^*|/|\mathcal{X}|$. Therefore, the expected number of samples needed to find a desired solution is $|\mathcal{X}|/|\mathcal{X}^*|$. Note that the target region

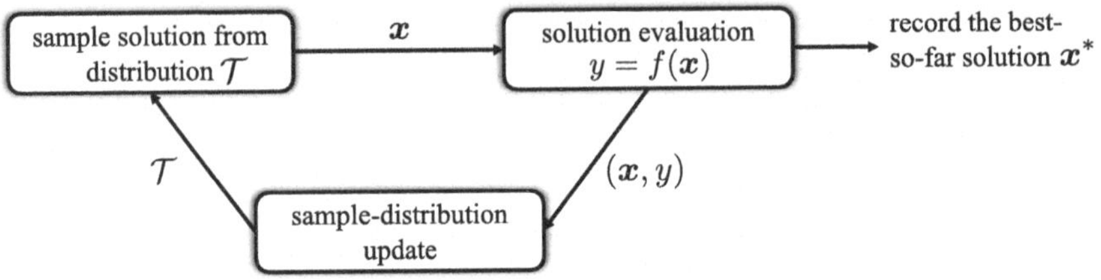

Fig. 1.1 A framework for derivative-free optimization algorithms

is usually very small, requiring many samples. For example, finding $(1, 1, \ldots, 1)$ in $\{0, 1\}^n$ by random search takes 2^n samples on average.

The inefficiency of random search stems from its blind sampling, as it does not learn from past samples. In other words, random search performs pure exploration. To improve efficiency, we can reduce exploration by adjusting the sampling distribution $\mathcal{T}$. As long as $\mathcal{T}$ assigns non-zero probability to any solution, the DFO method will ultimately converge to the global optimum. The question then becomes how to design a good sampling distribution $\mathcal{T}$. A well-designed sampling distribution is believed to accelerate convergence. However, the acceleration depends on the specific problem due to the No Free Lunch Theorem.

1.2.2 Development of DFO Algorithms

Various DFO algorithms have been proposed based on different ideas and inspirations. Early representative algorithms include simulated annealing (SA) [8], beam search [13], and evolutionary algorithms (EAs) [1]. SA generates new solutions by randomly perturbing the current solution and accepts them probabilistically based on the objective value difference and a decreasing temperature parameter. Beam search maintains a set of candidate solutions and iteratively expands and selects them based on their objective values. EAs are population-based algorithms inspired by natural evolution that generate new solutions using mutation and crossover operators and select the fittest ones to form the next population. These early DFO algorithms often employ heuristic rules to perturb solutions and generate new ones.

Later, estimation of distribution algorithms (EDAs) [9] were proposed, which explicitly model the distribution of promising solutions and sample new ones from the learned model. Bayesian optimization [15] builds a surrogate model to approximate the objective function and selects the next solution to evaluate by maximizing an acquisition function that balances exploration and exploitation.

DFO has also been an important research topic in the classical optimization community [3]. Notable methods include trust region methods [2], which approximate the objective function using a local model within a trust region, and pattern search methods [17], which perform exploratory searches along coordinate directions.

The variety of DFO algorithms stems from their different philosophies in balancing exploration and exploitation and their assumptions about the characteristics of the objective function. Understanding the properties and applicable scenarios of different algorithms is crucial for their effective use.

1.3 Automatic Machine Learning

The abundance of machine learning algorithms creates both opportunities and challenges. Powerful tools are available to tackle diverse learning tasks, but there is rarely a single algorithm that performs optimally on all tasks. The performance of a machine learning algorithm depends on factors such as the data, model architecture, and hyper-parameter settings. Choosing and configuring a suitable algorithm for a specific task often requires considerable expertise and experimentation.

Automatic machine learning (AutoML) aims to automate this process and enable non-experts to benefit from machine learning. A typical goal of AutoML is to find, for a given task, the combination of algorithm components and hyper-parameters that maximizes a predefined performance metric. This essentially treats AutoML as an optimization problem, with the solution space being the combinatorial space of algorithms and hyper-parameters, and the objective function being the performance metric.

However, evaluating the objective function in AutoML is often expensive, as it requires training and validating machine learning models. Moreover, the solution space is typically discrete, high-dimensional, and contains complex conditional dependencies. These properties make AutoML a challenging optimization problem that demands sample-efficient and flexible optimization algorithms.

DFO algorithms have found successful applications in AutoML. Bayesian optimization has been a popular choice due to its sample efficiency and ability to handle black-box objective functions [5, 16]. Evolutionary algorithms have also demonstrated competitive performance, especially for high-dimensional and conditional spaces [11, 14]. Recent studies showed promising results of combining Bayesian optimization with meta-learning to warm-start and guide the optimization on a new task [6].

Besides searching for performance-optimized models, another important aspect of AutoML is to automate the composition and configuration of machine learning pipelines, which include data preprocessing, feature engineering, model selection, and ensemble construction. This often requires optimizing multiple competing objectives, such as model performance, inference time, and complexity. Derivative-free multiobjective optimization algorithms have been adapted to solve such problems [7].

Furthermore, AutoML systems need to handle various practical issues such as computational resource constraints, noisy evaluations, and online adaptation to new data. Addressing these issues requires extending and enhancing DFO algorithms to meet the needs of AutoML, such as using multi-fidelity optimization to efficiently allocate resources [10].

The emergence of AutoML has brought new opportunities and challenges for DFO. It motivates the design of algorithms that can efficiently search high-dimensional, structured, and dynamic spaces. It also calls for rigorous benchmarking to evaluate and compare different DFO algorithms for AutoML. Research progress in these directions will make both AutoML and DFO more effective and widely applicable.

1.4 Organization of the Book

Aiming to build the theoretical foundations of DFO and design better optimization algorithms for machine learning, this book is organized into four parts, covering the preliminaries, analysis methodology, theoretical perspectives, and applications to AutoML.

Part I provides an introduction to DFO and its theoretical foundations. We present an overview of classic DFO algorithms, including direct search methods, model-based methods, and stochastic search methods. We also review the theoretical results on the approximation ability of DFO algorithms, highlighting the challenges and opportunities for theoretical analysis.

Part II presents a novel theoretical framework for analyzing and designing DFO algorithms. We introduce a general framework based on classification, which unifies existing algorithms and facilitates the design of new ones. We also present a basic algorithm RACOS designed under this framework.

Part III presents extensions of the RACOS algorithm. We design methods for more efficient optimization, high-dimensional problems, noisy problems, and large-scale optimizations. We also introduce a toolbox ZOOpt that includes the major algorithms in this book.

Part IV showcases the applications and integrations of DFO algorithms in automatic machine learning systems for hyper-parameter and algorithm selection.

Through this book, we aim to provide an alternative theoretical treatment of derivative-free optimization and present practical algorithmic innovations. We hope that the insights and techniques presented will inspire new research and empower practitioners to tackle real-world optimization challenges in machine learning and beyond.

References

1. Bäck T (1996) Evolutionary algorithms in theory and practice: evolution strategies, evolutionary programming, genetic algorithms. Oxford University Press, Oxford, UK
2. Conn AR, Gould NI, Toint PL (2000) Trust region methods
3. Conn AR, Scheinberg K, Vicente LN (2009) Introduction to derivative-free optimization, vol 8. Siam (2009)

4. Domingos P (2012) A few useful things to know about machine learning. Commun ACM 55(10):78–87
5. Feurer M, Klein A, Eggensperger K, Springenberg J, Blum M, Hutter F (2015) Efficient and robust automated machine learning. In: Advances in neural information processing systems, pp 2962–2970
6. Feurer M, Springenberg J, Hutter F (2015) Initializing bayesian hyperparameter optimization via meta-learning. In: Twenty-ninth AAAI conference on artificial intelligence
7. Horn D, Demircioğlu A, Bischl B, Glasmachers T, Weihs C (2016) A multi-objective auto-tuning framework for parallel coordinates plots. In: Proceedings of the 2016 on genetic and evolutionary computation conference, pp 1255–1262
8. Kirkpatrick S, Gelatt CD, Vecchi MP (1983) Optimization by simulated annealing. Science 220(4598):671–680
9. Larrañaga P, Lozano J (2002) Estimation of distribution algorithms: a new tool for evolutionary computation. Kluwer, Boston, MA
10. Li L, Jamieson K, DeSalvo G, Rostamizadeh A, Talwalkar A (2017) Hyperband: a novel bandit-based approach to hyperparameter optimization. J Mach Learn Res 18:6765–6816. JMLR.org
11. Liu H, Simonyan K, Vinyals O, Fernando C, Kavukcuoglu K (2020) Evolving deep neural networks. Artif Intell 288:103353
12. Mitchell T (1997) Machine learning. McGraw Hill, New York, NY
13. Ow PS, Morton TE (1988) Filtered beam search in scheduling. Int J Prod Res 26(1):35–62
14. Real E, Moore S, Selle A, Saxena S, Suematsu YL, Tan J, Le QV, Kurakin A (2017) Large-scale evolution of image classifiers. In: Proceedings of the 34th international conference on machine learning (ICML), pp 2902–2911, Sydney, Australia
15. Shahriari B, Swersky K, Wang Z, Adams RP, Freitas ND (2015) Taking the human out of the loop: a review of Bayesian optimization. Proc IEEE 104(1):148–175
16. Snoek J, Larochelle H, Adams RP (2012) Practical Bayesian optimization of machine learning algorithms. In: Advances in neural information processing systems 25, pp 2960–2968, Lake Tahoe, Nevada
17. Torczon V (1997) On the convergence of pattern search algorithms. SIAM J Optim 7(1):1–25

Chapter 2
Preliminaries

Abstract This chapter provides an overview of fundamental derivative-free optimization (DFO) algorithms, focusing on evolutionary algorithms (EAs), estimation of distribution algorithms (EDAs), and Bayesian optimization (BO). EAs are inspired by natural evolution. EDAs model the probability distribution of promising solutions to guide the search, while BO uses surrogate models to efficiently optimize expensive black-box functions. The chapter also discusses running time analysis, a key theoretical tool for understanding algorithm performance, and introduces the No Free Lunch Theorem, which highlights the importance of problem-specific knowledge in optimization. These concepts lay the groundwork for analyzing and designing advanced DFO methods, particularly in machine learning and other complex domains.

This chapter introduces some basic derivative-free optimization algorithms, including evolutionary algorithms (EAs), estimation of distribution algorithms (EDAs), and Bayesian optimization (BO). We will focus on representatives of EA for discrete optimization and evolutionary strategy (ES) for continuous optimization, and provide explanations of EDA and BO. These algorithms serve as the preliminary for understanding and analyzing more complex optimization methods. Additionally, we discuss some basic knowledge about the running time complexity of algorithms, which is the classical way to understand algorithms from a theoretical perspective.

2.1 Evolutionary Algorithms

Evolutionary algorithms (EAs) are a class of general-purpose heuristic optimization algorithms that simulate the natural evolution process by considering two key factors: variational reproduction and superior selection [1]. They repeatedly reproduce solutions by varying the currently maintained ones and eliminate inferior solutions, leading to iterative improvement. In this section, we introduce two representative population-based EAs: the $(\mu + \lambda)$-EA for discrete optimization and the $(\mu/\mu, \lambda)$-ES for continuous optimization.

Y. Yu et al., *Derivative-Free Optimization*, Machine Learning: Foundations, Methodologies, and Applications, https://doi.org/10.1007/978-981-96-5929-6_2

2.1.1 $(\mu + \lambda)$-EA

The $(\mu + \lambda)$-EA algorithm, presented in Algorithm 2.1, is a population-based EA for maximizing pseudo-Boolean functions over $\{0, 1\}^n$. The algorithm maintains a population of μ solutions and generates λ offspring solutions in each iteration. It starts with a randomly initialized population of μ solutions (line 1). In each iteration, λ offspring solutions are generated by applying a mutation operator to the parent solutions (lines 3–5). The mutation operator flips each bit of a solution independently with probability $1/n$. The μ best solutions are then selected from the union of the parent and offspring populations to form the next generation (line 6). This process continues until a stopping criterion is met.

Algorithm 2.1 $(\mu + \lambda)$-EA

Require: pseudo-Boolean function f : over$\{0, 1\}^n \rightarrow \mathbb{R}$, population size μ, offspring size λ
Ensure:
1: initialize a population P of μ solutions uniformly at random;
2: **while** stopping criterion is not met **do**
3:   let $P' = \emptyset$;
4:   **for** $i = 1$ to λ **do**
5:     generate $\boldsymbol{x}'$ by flipping each bit of a randomly selected solution $\boldsymbol{x} \in P$ independently with probability $1/n$; add $\boldsymbol{x}'$ to P';
6:   **end for**
7:   select the μ best solutions from $P \cup P'$ to form the next population P;
8: **end while**
9: **return** the best solution found

The $(\mu + \lambda)$-EA algorithm balances exploration and exploitation by adjusting the population size μ and the offspring size λ. A larger μ helps maintain diversity and enables global exploration, while a larger λ increases the selection pressure and focuses the search on promising regions. The optimal settings of μ and λ depend on the characteristics of the problem, such as the modality and the ruggedness of the fitness landscape.

2.1.2 $(\mu/\mu, \lambda)$-ES

The $(\mu/\mu, \lambda)$-ES algorithm, presented in Algorithm 2.2, is a population-based evolutionary strategy for continuous optimization problems. It maintains a population of μ solutions, each associated with a strategy parameter σ that controls the step size of the mutation operator. In each iteration, λ offspring solutions are generated by applying a mutation operator to the parent solutions (lines 4–6). The mutation operator adds a normally distributed random vector to each parent solution, where the random vector has zero mean and standard deviation σ in each dimension. The strategy parameter σ is also mutated by multiplying it with a log-normally distributed random factor (line 5). The μ best offspring solutions are then selected to form the next generation (line 7). This process continues until a stopping criterion is met.

Algorithm 2.2 $(\mu/\mu, \lambda)$-ES

Require: objective function $f : \mathbb{R}^n \rightarrow \mathbb{R}$, population size μ, offspring size λ, initial step size σ_0
Ensure:

1: initialize a population P of μ solutions $\boldsymbol{x}_i \in \mathbb{R}^n$ uniformly at random, and set $\sigma_i = \sigma_0$ for each solution;
2: **while** stopping criterion is not met **do**
3: let $P' = \emptyset$;
4: **for** $i = 1$ to λ **do**
5: select a parent solution $\boldsymbol{x} \in P$ uniformly at random, and mutate its strategy parameter $\sigma = \sigma \cdot \exp(\tau \cdot \mathcal{N}(0, 1))$, where $\tau = (\sqrt{2n})^{-1}$;
6: generate an offspring solution $\boldsymbol{x}' = \boldsymbol{x} + \sigma \cdot \mathcal{N}(\boldsymbol{0}, \boldsymbol{I})$, where $\mathcal{N}(\boldsymbol{0}, \boldsymbol{I})$ is a vector of independent standard normal random variables; add $(\boldsymbol{x}', \sigma)$ to P';
7: **end for**
8: select the μ best offspring solutions from P' to form the next population P;
9: **end while**
10: **return** the best solution found

The $(\mu/\mu, \lambda)$-ES algorithm adapts the step size of the mutation operator to the local characteristics of the objective function. The log-normal mutation of the strategy parameter σ allows the algorithm to self-adapt the step size based on the success of the previous mutations. If larger steps lead to better offspring solutions, the strategy parameter will increase, encouraging further exploration. If smaller steps are more successful, the strategy parameter will decrease, focusing the search on a smaller region. This self-adaptation mechanism enables the algorithm to efficiently explore the search space and converge to an optimal solution.

The $(\mu/\mu, \lambda)$-ES algorithm is a special case of the more general $(\mu/\rho, \lambda)$-ES, where ρ denotes the number of parent solutions used to generate each offspring solution. When $\rho = 1$, the algorithm is called a $(\mu/1, \lambda)$-ES or a (μ, λ)-ES. When $\rho = \mu$, the algorithm is called a $(\mu/\mu, \lambda)$-ES. The choice of ρ affects the balance between exploration and exploitation, with larger values of ρ promoting more exploitation.

The population-based EAs presented in this section have been widely studied in the theoretical analysis of evolutionary computation. They serve as a foundation for understanding the behavior and performance of more advanced EA variants and have inspired the design of effective optimization algorithms for various domains.

2.2 Estimation of Distribution Algorithms

Estimation of distribution algorithms (EDAs) are a class of EAs that explicitly model the probability distribution of promising solutions and sample new solutions from the learned model [9]. Instead of using traditional variation operators like mutation and crossover, EDAs build a statistical model of the selected solutions and generate new solutions by sampling from this model. The model is updated iteratively to reflect the distribution of increasingly better solutions.

The general procedure of an EDA is described in Algorithm 2.3. The algorithm starts with a randomly initialized population of solutions (line 1). In each iteration, a subset of promising solutions is selected from the current population according to a selection method (line 3). A probabilistic model is then built based on the selected solutions (line 4). This model captures the statistical dependencies among the variables of the selected solutions. New solutions are sampled from the learned model (line 5) and are used to replace some or all solutions in the population (line 6). This process continues until a stopping criterion is met.

Algorithm 2.3 Estimation of Distribution Algorithm (EDA)

Require: objective function f, population size N
Ensure:
1: initialize a population P of N solutions randomly;
2: **while** stopping criterion is not met **do**
3: select a subset S of promising solutions from P;
4: build a probabilistic model M based on the solutions in S;
5: sample a set O of new solutions from the model M;
6: replace some or all solutions in P with the solutions in O;
7: **end while**
8: **return** the best solution found

EDAs differ in the way they represent and learn the probabilistic model. Some common model representations include Bayesian networks, Markov networks, and multivariate normal distributions [10]. The choice of the model depends on the characteristics of the problem, such as the type of variables (discrete or continuous), the interactions among variables, and the complexity of the problem.

One of the most popular EDAs for discrete optimization is the Population-Based Incremental Learning (PBIL) algorithm [2]. In PBIL, the probabilistic model is represented as a vector of probabilities, where each element represents the probability of a variable taking the value 1 in the selected solutions. The model is initialized with a probability of 0.5 for each variable. In each iteration, the model is updated by moving the probability vector towards the best solution in the current population. New solutions are then sampled from the updated model by generating a binary vector according to the probabilities. The PBIL algorithm has been successfully applied to various combinatorial optimization problems, such as the traveling salesman problem and the knapsack problem [9].

For continuous optimization, the Covariance Matrix Adaptation Evolution Strategy (CMA-ES) is a well-known EDA [6]. In CMA-ES, the probabilistic model is a multivariate normal distribution, characterized by a mean vector and a covariance matrix. The mean vector represents the center of the distribution, while the covariance matrix captures the dependencies among the variables. In each iteration, a set of new solutions is sampled from the current distribution. The mean vector and the covariance matrix are then updated based on the best solutions in the sample. The update rules are designed to adapt the shape and the scale of the distribution to the local landscape of the objective function. CMA-ES has been shown to be highly effective on a wide range of continuous optimization problems [5].

2.3 Bayesian Optimization

Bayesian optimization (BO) is a model-based optimization approach that is particularly effective for expensive black-box functions [12]. The key idea of BO is to build a surrogate model of the objective function based on the observed data points and use this model to guide the search for the optimum. The surrogate model provides a probabilistic approximation of the objective function, allowing the algorithm to balance between exploring unknown regions and exploiting promising areas.

The general procedure of Bayesian optimization is described in Algorithm 2.4. The algorithm starts with an initial set of observations D, which contains the evaluated points and their corresponding objective values (line 1). A surrogate model M is then built based on the current observations (line 2). The most common choice for the surrogate model is the Gaussian process (GP) [11], which provides a probabilistic distribution over functions. A GP is specified by a mean function $m(\boldsymbol{x})$ and a covariance function $k(\boldsymbol{x}, \boldsymbol{x}')$, also known as the kernel. The mean function represents the expected value of the function at a given point, while the covariance function captures the similarity between two points.

Algorithm 2.4 Bayesian Optimization (BO)

Require: objective function f, initial observations D
Ensure:

1: initialize the observation set D with a few evaluated points;
2: **while** stopping criterion is not met **do**
3: build a surrogate model M based on the observations in D;
4: select the next point $\boldsymbol{x}_{next}$ to evaluate by optimizing an acquisition function $\alpha(\boldsymbol{x} \mid M)$;
5: evaluate the objective function at $\boldsymbol{x}_{next}$ to obtain $f(\boldsymbol{x}_{next})$;
6: add the new observation $(\boldsymbol{x}_{next}, f(\boldsymbol{x}_{next}))$ to D;
7: **end while**
8: **return** the best solution found

Given a set of observations, the GP posterior provides a normal distribution for the value of the function at any point. The mean and variance of this distribution can be used to guide the search for the optimum. Points with high mean values are considered promising and are more likely to be selected for evaluation (exploitation). Points with high variance values are considered uncertain and may lead to improvement if evaluated (exploration).

The selection of the next point to evaluate is performed by optimizing an acquisition function $\alpha(\boldsymbol{x} \mid M)$ (line 4). The acquisition function measures the expected benefit of evaluating the objective function at a given point, based on the current surrogate model. Some common acquisition functions include:

- Probability of Improvement (PI): $\alpha(\boldsymbol{x}) = \Phi((\mu(\boldsymbol{x}) - f(\boldsymbol{x}_{best}))/\sigma(\boldsymbol{x}))$, where $\Phi(\cdot)$ is the standard normal cumulative distribution function, $\mu(\boldsymbol{x})$ and $\sigma(\boldsymbol{x})$ are the mean and standard deviation of the GP posterior at $\boldsymbol{x}$, and $f(\boldsymbol{x}_{best})$ is the best objective value observed so far.

- Expected Improvement (EI): $\alpha(\boldsymbol{x}) = (\mu(\boldsymbol{x}) - f(\boldsymbol{x}_{best})) \cdot \Phi((\mu(\boldsymbol{x}) - f(\boldsymbol{x}_{best}))/\sigma(\boldsymbol{x})) + \sigma(\boldsymbol{x}) \cdot \phi((\mu(\boldsymbol{x}) - f(\boldsymbol{x}_{best}))/\sigma(\boldsymbol{x}))$, where $\phi(\cdot)$ is the standard normal probability density function.

- Upper Confidence Bound (UCB): $\alpha(\boldsymbol{x}) = \mu(\boldsymbol{x}) + \kappa \cdot \sigma(\boldsymbol{x})$, where κ is a parameter that controls the trade-off between exploration and exploitation.

The next point to evaluate is chosen as the one that maximizes the acquisition function (line 4). This point is then evaluated on the objective function, and the new observation is added to the data set (lines 5–6). The surrogate model is updated with the new observation, and the process is repeated until a stopping criterion is met.

Bayesian optimization has been successfully applied to a wide range of problems, including hyper-parameter tuning of machine learning algorithms [13], robot gait optimization [3], and chemical design [4]. The main advantage of BO is its sample efficiency, as it can find good solutions with a relatively small number of function evaluations. However, the performance of BO depends on the choice of the surrogate model and the acquisition function, as well as the dimensionality of the problem. In high-dimensional spaces, the number of observations required to build an accurate surrogate model increases exponentially, which limits the applicability of BO to problems with a moderate number of variables.

2.4 Running Time Analysis

The running time of an algorithm is a measure of its computational complexity, which is usually expressed as a function of the input size. In the context of EAs, the running time is often analyzed in terms of the expected number of function evaluations needed to find the optimal solution or to achieve a certain approximation ratio.

One common approach to running time analysis is to model the EA as a Markov chain [7]. A Markov chain is a stochastic process that transitions between states, where the transition probabilities depend only on the current state. By defining the states of the Markov chain as the populations of the EA and the transition probabilities based on the selection and variation operators, we can analyze the expected time for the EA to reach a desired state, such as the global optimum.

Two classical methods for running time analysis of EAs modeled as Markov chains are the fitness level method [14], the drift analysis [8], and the convergence-based analysis [16]. The fitness level method partitions the search space into a sequence of fitness levels and estimates the expected time to progress from one level to the next. The total running time is then the sum of the expected times over all levels. Drift analysis measures the expected progress towards the optimal solution in each iteration and uses this to bound the total running time. Convergence-based analysis bridges the convergence rate and the running time complexity. The running time complexity can thus be derived from the convergence rate. The switch analysis [17] is an advanced analysis approach that can unify these three approaches. More comprehensive introduction of these analysis approaches can be found in [18].

However, running time analysis of EAs is still a challenging task, especially for complex problems and advanced EA variants, which is the same for other DFO methods. Techniques from probability theory, combinatorics, and graph theory are often employed to derive tight bounds on the running time.

2.5 No Free Lunch in Optimization

When considering general-purpose optimization techniques, the No Free Lunch (NFL) Theorem is a fundamental principle that should never be overlooked.

The NFL Theorem, formulated by Wolpert and Macready [15], states that no optimization algorithm is universally better than any other when their performance is averaged across all possible problems. Specifically, the NFL theorem assumes that the algorithm has no prior knowledge about the objective function. Under this assumption, if an algorithm outperforms random search on some problems, it must underperform on others when uniformly averaged over all possible objective functions.

Mathematically, let $\mathcal{X}$ be a finite solution space, $\mathcal{Y}$ a finite set of objective values, and $\mathcal{F}$ the set of all possible objective functions $f : \mathcal{X} \rightarrow \mathcal{Y}$. Let $d(\cdot, \cdot)$ be a performance measure, such as the best function value found within a fixed number of iterations. The NFL Theorem states that for any two algorithms a_1 and a_2,

$$\sum_{f \in \mathcal{F}} d(a_1, f) = \sum_{f \in \mathcal{F}} d(a_2, f). \tag{2.1}$$

An intuitive proof is given below.

Proof Consider the set of all possible objective functions $f : \mathcal{X} \rightarrow \mathcal{Y}$, where the algorithm has no prior knowledge. For two algorithms A_1 and A_2, define a performance measure $d(A, f)$ that quantifies how well A performs on f.

Fix a sequence of m solutions that A_1 evaluates for a given f. There are $|\mathcal{Y}|^m$ possible ways to assign objective values to these solutions. Among these, some assignments favor A_1, while others do not.

Now consider A_2. For each assignment favoring A_1, there exists a corresponding assignment favoring A_2 to the same extent, as A_2 could simply evaluate the solutions in a different order.

Therefore, for every objective function where A_1 outperforms A_2, there exists another where A_2 outperforms A_1 by the same margin. Averaged over all possible objective functions, their performance is equal.

This argument holds for any two algorithms and any performance measure. Thus, without prior knowledge, no algorithm can outperform another when averaged over all possible functions. □

The NFL Theorem implies that the performance of any two algorithms is equal when averaged over all possible objective functions. It reveals the fundamental difficulty in designing general-purpose optimization algorithms—prior knowledge is necessary for an algorithm to outperform random search. Without prior knowledge, an algorithm can only excel on a subset of problems at the cost of inferior performance on others.

This leads to a crucial question: for a given optimization algorithm, on which subset of problems does it excel? Identifying the characteristics of problems that an algorithm is well-suited for is one of the most significant considerations in the design and analysis of DFO algorithms. Understanding an algorithm's strengths and limitations allows researchers and practitioners to make informed decisions when selecting or designing optimization techniques for specific applications.

References

1. Bäck T (1996) Evolutionary algorithms in theory and practice: evolution strategies, evolutionary programming, genetic algorithms. Oxford University Press, Oxford, UK
2. Baluja S (1994) Population-based incremental learning. a method for integrating genetic search based function optimization and competitive learning. Carnegie-Mellon Univ Pittsburgh Pa Dept Of Computer Science
3. Calandra R, Seyfarth A, Peters J, Deisenroth MP (2016) Bayesian optimization for learning gaits under uncertainty. In: Annals of mathematics and artificial intelligence, vol 76. Springer, pp 5–23
4. Griffiths R-R, Hernández-Lobato JM (2020) Constrained bayesian optimization for automatic chemical design using variational autoencoders. Chem Sci 11(2):577–586
5. Hansen N (2006) The cma evolution strategy: a comparing review. Towards a new evolutionary computation, pp 75–102
6. Hansen N, Ostermeier A (2001) Completely derandomized self-adaptation in evolution strategies. Evol Comput 9(2):159–195
7. He J, Yao X (2003) Towards an analytic framework for analysing the computation time of evolutionary algorithms. Artif Intell 145(1–2):59–97
8. He J, Yao X (2004) A study of drift analysis for estimating computation time of evolutionary algorithms. Nat Comput 3(1):21–35
9. Larrañaga P, Lozano J (2002) Estimation of distribution algorithms: a new tool for evolutionary computation. Kluwer, Boston, MA
10. Pelikan M, Goldberg DE, Cantu-Paz E (2002) A survey of optimization by building and using probabilistic models. Comput Optim Appl 21(1):5–20
11. Rasmussen CE (2003) Gaussian processes in machine learning. Summer school on machine learning, pp 63–71
12. Shahriari B, Swersky K, Wang Z, Adams RP, Freitas ND (2015) Taking the human out of the loop: a review of Bayesian optimization. Proc IEEE 104(1):148–175
13. Snoek J, Larochelle H, Adams RP (2012) Practical Bayesian optimization of machine learning algorithms. In: Advances in neural information processing systems, vol 25. Lake Tahoe, Nevada, pp 2960–2968

14. Wegener I (2002) Methods for the analysis of evolutionary algorithms on pseudo-boolean functions. Evolutionary optimization, pp 349–369
15. Wolpert D, Macready W (1997) No free lunch theorems for optimization. IEEE Trans Evol Comput 1(1):67–82
16. Yu Y, Zhou Z-H (2008) A new approach to estimating the expected first hitting time of evolutionary algorithms. Artif Intell 172(15):1809–1832
17. Yu Y, Qian C, Zhou Z-H (2015) Switch analysis for running time analysis of evolutionary algorithms. IEEE Trans Evol Comput 19(6):777–792
18. Zhou Z-H, Yu Y, Qian C (2019) evolutionary learning: advances in theories and algorithms. Springer, Berlin. ISSN 978-981-13-5955-2

Part II
Classification-Based Derivative-Free Optimization

Chapter 3
Framework

Abstract This chapter introduces the Sampling-and-Learning (SAL) framework, a unifying approach to understanding derivative-free optimization (DFO) algorithms. The SAL framework consists of two main stages: sampling, where new candidate solutions are generated, and learning, where promising solutions are selected to guide future sampling. The framework iteratively alternates between these stages, using a learned model to represent the algorithm's belief about promising regions in the solution space. The chapter also presents a simplified version called the Sampling-and-Classification (SAC) framework, which uses binary classification to distinguish between promising and unpromising solutions. The SAL and SAC frameworks provide a statistical perspective on how DFO algorithms balance exploration and exploitation, offering a systematic way to analyze and design optimization methods. The chapter concludes by discussing the challenges and potential of these frameworks, including the computational overhead of learning and the need for accurate models to guide the search effectively.

In the previous chapter, we explored a variety of derivative-free optimization algorithms, including evolutionary algorithms, estimation of distribution algorithms, and Bayesian optimization. These algorithms are particularly effective for solving complex optimization problems where direct analytical evaluation of the objective function is either infeasible or too computationally expensive. They have been successfully applied in diverse fields where objective functions remain hidden or are too intricate for traditional gradient-based methods.

Upon closer examination, the reader may have observed a recurring theme or underlying principle across these seemingly different algorithms. Despite their distinct methodologies, these algorithms appear to share a common strategy or conceptual foundation. In this chapter, we aim to formalize this observation by introducing a unifying framework known as the *sampling-and-learning* (SAL) framework [5, 6]. This framework provides a statistical lens through which we can better understand the mechanisms that drive the success of these heuristics. By examining these algorithms from a statistical perspective, the SAL framework helps explain how they effectively balance exploration and exploitation, ultimately guiding the search process toward optimal solutions.

Y. Yu et al., *Derivative-Free Optimization*, Machine Learning: Foundations, Methodologies, and Applications, https://doi.org/10.1007/978-981-96-5929-6_3

3.1 Sampling and Learning Framework

The SAL framework consists of two main stages: the sampling stage, which models the generation of new candidate solutions, and the learning stage, which models the selection of promising solutions to guide the next sampling step. The framework starts with a randomly initialized set of solutions and then iteratively alternates between the sampling and learning stages. In the sampling stage, new solutions are generated based on the current learned model, which represents the algorithm's belief about the location of promising solutions. In the learning stage, the generated solutions are evaluated, and the learned model is updated based on the obtained information. This process continues until a satisfactory solution is found or a predetermined termination criterion is met.

Formally, let X denote the solution space, and let $f : X \to \mathbb{R}$ be the objective function to be minimized. The SAL framework maintains a set of candidate solutions S_t at each iteration t. The learning stage at iteration t consists of constructing a learned model h_t based on the current solution set S_{t-1} and their corresponding objective function values. The model h_t represents the algorithm's belief about the location of promising solutions in the solution space. In the sampling stage, a new set of candidate solutions S_t is generated by sampling from a distribution $\mathcal{T}_{h_t}$ induced by the learned model h_t.

The learning stage can be formally described as follows:

$$h_t = \mathcal{L}(S_{t-1}, f(S_{t-1}), h_{t-1}), \tag{3.1}$$

where $\mathcal{L}$ is the learning algorithm, S_{t-1} is the set of candidate solutions from the previous iteration, $f(S_{t-1})$ are their corresponding objective function values, and h_{t-1} is the learned model from the previous iteration. The learning algorithm $\mathcal{L}$ can be any suitable machine learning technique, such as regression, classification, or density estimation, depending on the specific optimization algorithm being modeled.

The sampling stage can be formally described as follows:

$$S_t = \{x_i \sim \mathcal{T}_{h_t} \mid i = 1, \ldots, m_t\}, \tag{3.2}$$

where $\mathcal{T}_{h_t}$ is the sampling distribution induced by the learned model h_t, and m_t is the number of candidate solutions to be generated at iteration t. The sampling distribution $\mathcal{T}_{h_t}$ reflects the algorithm's belief about the location of promising solutions and is used to guide the search towards regions of the solution space that are likely to contain good solutions.

The SAL framework can be summarized in the following algorithm:

The SAL framework provides a general template for modeling and analyzing a wide range of derivative-free optimization algorithms. By specifying different learning algorithms $\mathcal{L}$ and sampling distribution transformers $\mathcal{T}$, the framework can capture the behavior of various algorithms, such as evolutionary algorithms, estimation of distribution algorithms, and Bayesian optimization.

Algorithm 3.1 Sampling and Learning (SAL) Framework

Require: Objective function f, learning algorithm $\mathcal{L}$, sampling distribution transformer $\mathcal{T}$, number of iterations T, initial solution set S_0
Ensure: Best solution found
1: **for** $t = 1, \dots, T$ **do**
2: Construct the learned model $h_t = \mathcal{L}(S_{t-1}, f(S_{t-1}), h_{t-1})$
3: Generate a new set of candidate solutions $S_t = \{x_i \sim \mathcal{T}_{h_t} \mid i = 1, \dots, m_t\}$
4: Evaluate the objective function f for each solution in S_t
5: **end for**
6: **return** Best solution found in $\bigcup_{t=0}^{T} S_t$

3.2 Casting Previous DFO Methods Into the SAL Framework

The SAL framework provides a unifying perspective on various derivative-free optimization methods. By identifying the key components of these methods and mapping them to the sampling and learning stages of the SAL framework, we can gain insights into their underlying principles and develop a more systematic understanding of their behavior. In this section, we will discuss how some popular DFO methods, namely evolutionary algorithms [2], estimation of distribution algorithms [4], and Bayesian optimization, can be cast into the SAL framework.

3.2.1 *Estimation of Distribution Algorithms*

Estimation of distribution algorithms (EDAs) are a class of optimization algorithms that explicitly build a probabilistic model of promising solutions and use this model to generate new candidate solutions. EDAs can be naturally cast into the SAL framework as follows:

- **Sampling stage:** In EDAs, the sampling stage involves generating new candidate solutions by sampling from the learned probabilistic model. The probabilistic model captures the distribution of promising solutions in the solution space and guides the search towards regions of high probability. The specific sampling mechanism depends on the type of probabilistic model employed, such as Gaussian models, Bayesian networks, or Markov random fields.
- **Learning stage:** The learning stage in EDAs consists of building or updating the probabilistic model based on the selected solutions from the previous iteration. The selected solutions are typically the best-performing ones according to the objective function. The probabilistic model is learned using statistical learning techniques, such as maximum likelihood estimation, Bayesian inference, or graphical model learning. The learned model represents the algorithm's belief about the distribution of promising solutions.

- **Learned model:** In EDAs, the learned model is explicitly represented as a probabilistic model, such as a Gaussian distribution, a Bayesian network, or a Markov random field. The probabilistic model captures the dependencies and relationships among the variables of the optimization problem and provides a compact representation of the promising regions in the solution space. The learned model is used to guide the sampling stage in the next iteration.

3.2.2 Bayesian Optimization

Bayesian optimization (BO) is a class of optimization algorithms that build a surrogate model of the objective function and use this model to guide the search for optimal solutions. BO algorithms can be cast into the SAL framework as follows:

- **Sampling stage:** In BO, the sampling stage involves selecting the next point to evaluate based on the surrogate model and an acquisition function. The acquisition function balances exploration and exploitation by considering both the predicted performance and the uncertainty of the surrogate model. Common acquisition functions include expected improvement, probability of improvement, and upper confidence bound. The selected point is then evaluated using the expensive objective function.
- **Learning stage:** The learning stage in BO consists of updating the surrogate model based on the observed data points, i.e., the evaluated solutions and their corresponding objective function values. The surrogate model is typically a probabilistic model, such as a Gaussian process, that provides a probabilistic estimate of the objective function. The model is updated using Bayesian inference techniques, such as maximum likelihood estimation or Markov chain Monte Carlo (MCMC) methods.
- **Learned model:** In BO, the learned model is the surrogate model, which is a probabilistic representation of the objective function. The surrogate model captures the knowledge about the function's behavior based on the observed data points. It provides a probabilistic estimate of the function value at any point in the solution space, along with an associated uncertainty. The learned model is used to guide the sampling stage by informing the acquisition function.

3.2.3 Evolutionary Algorithms

Evolutionary algorithms (EAs) are a class of optimization algorithms inspired by the principles of natural evolution. Classical EAs maintain a population of candidate solutions and iteratively evolve the population through selection, reproduction, and variation operators. They do not maintain explicitly models of distribution. However, it should be noticed that the distribution can be equivalently expressed by the population of solutions. Likewise, in machine learning community, there is similarly a branch of case-based learning algorithms that maintains a set of samples, such as k-nearest neighbor classifiers. Therefore, the main components of EAs, such as genetic algorithms (GAs), evolution strategies (ES), and differential evolution (DE), can be mapped to the SAL framework as follows:

- **Sampling stage:** In EAs, the sampling stage corresponds to the generation of new candidate solutions through variation operators, such as mutation and crossover. These operators introduce diversity into the population and explore the solution space. For example, in GAs, mutation randomly modifies individual elements of a solution, while crossover combines genetic material from parent solutions to create offspring. The specific choice of variation operators depends on the representation of the solutions and the problem domain.
- **Learning stage:** The learning stage in EAs is represented by the selection operator, which chooses promising solutions from the current population to form the next generation. Selection operators assign higher probabilities to solutions with better fitness values, thus guiding the search towards promising regions of the solution space. Common selection schemes include tournament selection, roulette wheel selection, and truncation selection. The selected solutions serve as the basis for the next iteration of the algorithm.
- **Learned model:** In the context of EAs, the learned model can be interpreted as the distribution of the selected solutions in the solution space. The selection operator implicitly learns the characteristics of promising solutions by favoring those with higher fitness values. The distribution of the selected solutions represents the algorithm's belief about the location of good solutions and guides the sampling stage in the next iteration.

3.2.4 Other DFO Methods

The SAL framework can also accommodate other derivative-free optimization methods, such as Ant Colony Optimization [1], Particle Swarm Optimization [3], etc. These methods can be cast into the SAL framework by identifying the key components of sampling and learning:

- **Sampling stage:** The sampling stage in these methods involves generating new candidate solutions based on a set of predefined rules or heuristics. For example, pattern search methods explore the solution space by generating points along coordinate directions, while simplex methods create new points by reflecting, expanding, or contracting the current simplex.
- **Learning stage:** The learning stage in these methods typically involves updating the internal parameters or the search strategy based on the observed function values.

For example, trust-region methods adjust the size of the trust region based on the agreement between the surrogate model and the actual function values.

- **Learned model:** The learned model in these methods may not be as explicit as in EDAs or BO, but it still represents the algorithm's belief about the promising regions of the solution space. For example, in pattern search methods, the learned model can be considered as the set of search directions that have led to successful steps in previous iterations.

By mapping the key components of these methods to the sampling and learning stages, the SAL framework provides a common language to analyze and compare different DFO methods. We can understand the optimization power of DFO in a unified way. It also facilitates the development of hybrid approaches that combine the strengths of various methods. We can further leverage the theoretical foundations and analysis techniques developed for the framework to gain insights into the convergence properties and performance of these methods. This unified perspective helps in identifying the key factors that influence the effectiveness of DFO methods and guides the design of more efficient and principled optimization algorithms.

3.3 Sampling and Classification Framework

In addition to the general SAL framework, we also introduce a simplified version called the *sampling-and-classification* (SAC) framework. Before that, let's firstly re-design the SAL framework specifically in Algorithm 3.2

Algorithm 3.2 The sampling-and-learning (SAL) framework

Require:

$\alpha^* > 0$: Approximation level
$T \in \mathbb{N}^+$: Number of iterations
$m_0, \ldots, m_T \in \mathbb{N}^+$: Number of samples
$\lambda \in [0, 1]$: Balancing parameter
$\mathcal{L}$: Learning algorithm
$\mathcal{T}$: Distribution transformation of hypothesis

Ensure:

Collect $S_0 = \{\boldsymbol{x}_1, \ldots, \boldsymbol{x}_{m_0}\}$ by i.i.d. sampling from $\mathcal{U}_X$
$\tilde{\boldsymbol{x}} = \arg\min_{\boldsymbol{x} \in S_0} f(\boldsymbol{x})$
for $t = 1$ to T **do**
  Learn $h_t = \mathcal{L}(S_{t-1}, f(S_{t-1}), h_{t-1}, t)$
  for $i = 1$ to m_t **do**
    Sample $\boldsymbol{x}_i$ from $\begin{cases} \mathcal{T}_{h_t}, & \text{with probability } \lambda \\ \mathcal{U}_X, & \text{with probability } 1 - \lambda \end{cases}$
    $S_t = S_t \cup \{\boldsymbol{x}_i\}$
  end for
  $\tilde{\boldsymbol{x}} = \arg\min_{\boldsymbol{x} \in S_t \cup \{\tilde{\boldsymbol{x}}\}} f(\boldsymbol{x})$
end for
return $\tilde{\boldsymbol{x}}$

The SAL framework starts with random sampling in Step 1, like all derivative-free optimization methods. Steps 2 and 9 record the best-so-far solutions throughout the search. It follows a cycle consisting of learning and sampling stages. In Step 4, it learns a hypothesis h_t (i.e., a mapping from X to $\mathbb{R}$) via the learning algorithm $\mathcal{L}$. The learning algorithm allows taking the latest samples S_{t-1}, the labels $f(S_{t-1})$, the last hypothesis h_{t-1}, and the iteration time t into account. Different derivative-free optimization methods may make different use of them. In Steps 5 to 8, it samples from the distribution transformed from the hypothesis and the whole solution space, balanced by a probability, where the sample set S_t is initialized to be empty by default. The distribution $\mathcal{T}_{h_t}$ implies the potentially good regions learned by h_t.

Comparing with the framework in Algorithm 3.1, Algorithm 3.2 specifies the sampling distribution to be combined with the uniform distribution and the transformed distribution from the model. Note that this design reflects the exploration-exploitation balance in many DFO methods, and the uniform sampling ensures that the algorithm will converge to optimal solutions.

Now we can see that the SAL framework still leaves the learning algorithm and the model transformed distribution unspecified. We then choose to implement the learning algorithm to be classification algorithms, which result in classification models that lead to simple understanding and analysis of the framework as we will present in later chapters.

The sampling-and-classification (SAC) framework is a special case of the SAL framework where the learning stage employs a binary classification model to distinguish between promising and unpromising solutions. The classifier is trained on the evaluated solutions, labeling them as positive (promising) or negative (unpromising) based on a predefined threshold. The sampling stage then generates new solutions by focusing on the regions classified as promising by the learned model.

Formally, let $h_t : X \to \{-1, +1\}$ be a binary classifier learned at iteration t, where $h_t(x) = +1$ indicates that solution x is predicted to be promising, and $h_t(x) = -1$ indicates that x is predicted to be unpromising. The learning stage in the SAC framework can be described as follows:

$$h_t = \mathcal{C}(\{(x, \mathbb{I}[f(x) \le \alpha_t]) \mid x \in S_{t-1}\}), \tag{3.3}$$

where $\mathcal{C}$ is a binary classification algorithm, $\mathbb{I}[\cdot]$ is the indicator function, and α_t is a predefined threshold value at iteration t. The classifier h_t is trained on the solution set S_{t-1} from the previous iteration, with each solution labeled as positive if its objective function value is below the threshold α_t and negative otherwise.

The sampling stage in the SAC framework focuses on generating new solutions from the regions classified as promising by the learned classifier. This can be achieved by sampling from a distribution that assigns higher probabilities to solutions in the positive region of the classifier. One common approach is to use a truncated uniform distribution over the positive region:

$$\mathcal{T}_{h_t}(x) \propto \begin{cases} \mathcal{U}(x) & \text{if } h_t(x) = +1 \\ 0 & \text{otherwise} \end{cases}, \tag{3.4}$$

where $\mathcal{U}$ is the uniform distribution over the solution space X.

The implementation of the SAC framework only differs from the SAL framework in the learning algorithm, as summarized in Algorithm 3.3

Algorithm 3.3 The sampling-and-classification (SAC) framework

Require:

$\alpha^* > 0$: Approximation level
$T \in \mathbb{N}^+$: Number of iterations
$m_0, \ldots, m_T \in \mathbb{N}^+$: Number of samples
$\alpha_1, \ldots, \alpha_T$: Level parameters
$\lambda \in [0, 1]$: Balancing parameter
$\mathcal{L}$: Learning algorithm
$\mathcal{T}$: Distribution transformation of hypothesis

Ensure:

Collect $S_0 = \{\boldsymbol{x}_1, \ldots, \boldsymbol{x}_{m_0}\}$ by i.i.d. sampling from $\mathcal{U}_X$
$\tilde{\boldsymbol{x}} = \arg\min_{\boldsymbol{x} \in S_0} f(\boldsymbol{x})$
for $t = 1$ to T **do**
  Learn a binary classifier $h_t = \mathcal{C}(\{(x, \mathbb{I}[f(x) \leq \alpha_t]) \mid x \in S_{t-1}\})$
  for $i = 1$ to m_t **do**
    Sample $\boldsymbol{x}_i$ from $\begin{cases} \mathcal{T}_{h_t}, & \text{with probability } \lambda \\ \mathcal{U}_X, & \text{with probability } 1 - \lambda \end{cases}$
    $S_t = S_t \cup \{\boldsymbol{x}_i\}$
  end for
  $\tilde{\boldsymbol{x}} = \arg\min_{\boldsymbol{x} \in S_t \cup \{\tilde{\boldsymbol{x}}\}} f(\boldsymbol{x})$
end for
return $\tilde{\boldsymbol{x}}$

The effectiveness of the SAC framework depends on several factors, such as the choice of the classification algorithm, the selection of the threshold values, and the sampling strategy. In the next chapter, we will identify the key factors that influence its efficiency and provide guidelines for designing effective SAC algorithms.

3.4 Summary

In this chapter, we introduced the sampling-and-learning (SAL) and sampling-and-classification (SAC) frameworks as unifying principles behind various derivative-free optimization algorithms. The SAL framework provides a general template for modeling and analyzing algorithms that alternate between a sampling stage and a learning stage. The SAC framework is a simplified version of the SAL framework that employs a binary classification model to guide the search process.

The SAL and SAC frameworks offer a principled way to design and analyze derivative-free optimization algorithms. By formalizing the common strategies employed by these algorithms, the frameworks provide a deeper understanding of their mechanisms and allow for a more systematic analysis of their performance. The frameworks also open up new avenues for designing novel optimization algorithms by exploring different learning algorithms, sampling strategies, and hybridization techniques.

Meanwhile, the SAL and SAC frameworks also pose several challenges. The performance of these frameworks heavily depends on the quality of the learned models and the effectiveness of the sampling strategies. If the learned models are not accurate or do not capture the relevant features of the optimization problem, the search process may be misguided, leading to poor performance. Similarly, if the sampling strategies do not effectively explore the solution space or focus on the promising regions, the algorithm may get stuck in suboptimal solutions.

Another challenge is the computational overhead introduced by the learning component. Training the models and generating samples from complex distributions can be computationally expensive, especially in high-dimensional solution spaces. Balancing the cost of learning with the cost of function evaluations is crucial for the overall efficiency of the algorithm.

In the next chapter, we will delve into the theoretical foundations of the SAL and SAC frameworks. We will analyze their convergence properties, derive performance bounds, and identify the key factors that influence their efficiency. Through this theoretical investigation, we aim to provide a rigorous understanding of the strengths and limitations of these frameworks and guide the design of new derivative-free optimization algorithms.

References

1. Dorigo M, Maniezzo V, Colorni A (1996) Ant system: Optimization by a colony of cooperating agents. IEEE Trans. Syst. Man Cybern.–Part B **26**(1), 29–41 (1996)
2. Goldberg D (1989) Genetic algorithms in search, optimization, and machine learning. Addison-Wesley, Reading, MA
3. Kennedy J, Eberhart R (1995) Particle swarm optimization. In: Proceedings of the IEEE international conference on neural networks, vol 4. Perth, Australia, pp 1942–1948
4. Larrañaga P, Lozano J (2002) Estimation of distribution algorithms: a new tool for evolutionary computation. Kluwer, Boston, MA
5. Yu Y, Qian H (2014) The sampling-and-learning framework: a statistical view of evolutionary algorithm. In: Proceedings of the 2014 IEEE congress on evolutionary computation. Beijing, China, pp 149–158
6. Yu Y, Qian H, Hu Y-Q (2016) Derivative-free optimization via classification. In: Proceedings of the 30th AAAI conference on artificial intelligence, Phoenix, AZ, pp 2286–2292

Chapter 4
Theoretical Foundation

Abstract This chapter provides a theoretical foundation for the Sampling-and-Learning (SAL) and Sampling-and-Classification (SAC) frameworks in derivative-free optimization (DFO). It introduces the concept of (ϵ, δ)-query complexity, which measures the number of function evaluations required to find an ϵ-optimal solution with probability $1 - \delta$. The chapter derives general performance bounds for the SAL framework and identifies two key factors influencing the SAC framework's efficiency: error-target dependence and shrinking rate. These factors measure the alignment between the learned model and the target solution set, and the reduction in the search space volume, respectively. The analysis shows that SAC algorithms can achieve polynomial query complexity for functions with local Lipschitz continuity and bounded packing/covering numbers. The chapter concludes by discussing practical implications for designing efficient DFO algorithms, emphasizing the importance of model alignment and problem geometry in optimization performance.

Despite the success of DFO algorithms in practice, their theoretical foundations have been relatively less explored compared to their gradient-based counterparts. Rigorous theoretical analysis is crucial for understanding the convergence properties, sample complexity, and performance guarantees of these algorithms. It provides insights into the key factors that influence their efficiency and guides the design of more principled and effective optimization methods.

In the previous chapter, we introduced the sampling-and-learning (SAL) and sampling-and-classification (SAC) frameworks as unifying principles behind various DFO algorithms. These frameworks provide a general template for modeling and analyzing algorithms that alternate between a sampling stage and a learning stage. The SAL framework encompasses a wide range of DFO algorithms, while the SAC framework focuses on algorithms that employ a binary classification model to guide the search process.

The SAL and SAC frameworks offer a principled way to design and analyze DFO algorithms by formalizing the common strategies employed by these algorithms. They capture the essence of the exploration-exploitation trade-off, which is fundamental to the success of any optimization algorithm. By leveraging the tools and techniques from statistical learning theory and probability theory, we can derive performance bounds and convergence guarantees for these frameworks.

Y. Yu et al., *Derivative-Free Optimization*, Machine Learning: Foundations, Methodologies, and Applications, https://doi.org/10.1007/978-981-96-5929-6_4

In this chapter, we present a rigorous theoretical analysis of the SAL and SAC frameworks, focusing on their convergence properties, sample complexity, and the key factors influencing their efficiency. We introduce the concept of (ϵ, δ)-query complexity, which measures the number of function evaluations required by an algorithm to find an ϵ-optimal solution with probability at least $1 - \delta$. This complexity measure provides a unified way to compare the performance of different DFO algorithms and understand their scalability with respect to the problem dimension and the desired accuracy.

We start by deriving a general performance bound for the SAL framework in terms of the average success probability of sampling from the learned models. This bound highlights the importance of the learning stage in guiding the search towards promising regions of the solution space. We then specialize the analysis to the SAC framework and identify two key factors that influence its performance: the error-target dependence and the shrinking rate. The error-target dependence measures the alignment between the classification error and the target ϵ-optimal set, while the shrinking rate quantifies the reduction in the volume of the promising region across iterations.

To illustrate the applicability of our theoretical results, we consider two classes of objective functions: functions satisfying the local Lipschitz condition and functions with bounded packing and covering numbers. For functions satisfying the local Lipschitz condition, we show that the SAC framework can achieve a polynomial query complexity if the error-target dependence is strictly less than one and the shrinking rate is positive. For functions with bounded packing and covering numbers, we derive a sufficient condition under which the SAC framework achieves a polynomial query complexity.

4.1 Problem Setting and Notations

We consider general minimization problems.

Definition 4.1 (*Minimization Problem*) A minimization problem consists of a solution space X and a function $f : X \to \mathbb{R}$. The goal is to find a solution $\boldsymbol{x}^* \in X$ such that $f(\boldsymbol{x}^*) \leq f(\boldsymbol{x})$ for all $\boldsymbol{x} \in X$.

We assume without loss of generality that the value of f is bounded in $[0, 1]$, i.e., $\forall \boldsymbol{x} \in X \colon f(\boldsymbol{x}) \in [0, 1]$. Given an arbitrary function g with a bounded value range over the input domain, the bound can be implemented by a simple normalization

$$f(\boldsymbol{x}) = \frac{g(\boldsymbol{x}) - g(\boldsymbol{x}^*)}{\max_{\boldsymbol{x}'} g(\boldsymbol{x}') - g(\boldsymbol{x}^*)}.$$

Therefore, under our assumption, the optimal value is 0.

Throughout this chapter, we use the following notations:

- $\mathcal{A}$: an optimization algorithm.
- $\mathcal{D}$: a probability distribution over the solution space X.
- $\mathcal{U}$: the uniform distribution over X.
- $\mathcal{H}$: a hypothesis space, where each hypothesis $h \in \mathcal{H}$ is a function mapping from X to $\{-1, +1\}$.
- D_h: the region of X where a hypothesis $h \in \mathcal{H}$ predicts positive, i.e., $D_h = \{\boldsymbol{x} \in X \mid h(\boldsymbol{x}) = +1\}$.
- D_α: the α-sublevel set of f, i.e., $D_\alpha = \{\boldsymbol{x} \in X \mid f(\boldsymbol{x}) \leq \alpha\}$.
- D_ϵ: the ϵ-optimal set, i.e., $D_\epsilon = \{\boldsymbol{x} \in X \mid f(\boldsymbol{x}) - f(\boldsymbol{x}^*) \leq \epsilon\}$.
- $\mu(A) = |A|/|X|$: the ratio of set A taking place in X.
- $\Pr(\cdot)$: the probability of an event.
- $\mathbb{E}[\cdot]$: the expectation of a random variable.
- $\mathtt{poly}(\cdots)$: represents the set of all polynomials of the related variables.
- $\mathtt{superpoly}(\cdots)$: represents the set of all functions that grow faster than any function in $\mathtt{poly}(\cdots)$ with the related variables.

4.2 (ϵ, δ)-Query Complexity

To analyze the performance of the SAL and SAC frameworks, we introduce the concept of (ϵ, δ)-query complexity. Given an algorithm $\mathcal{A}$, the (ϵ, δ)-query complexity is the number of objective function evaluations required by $\mathcal{A}$ to find an ϵ-optimal solution with probability at least $1 - \delta$. This complexity measure reflects our belief that DFO methods are, instead of accurate solvers, powerful *approximate* solvers. They can produce good results in many, but not all, cases. It also reflects our adherence to the No Free Lunch theorem.

Definition 4.2 (*(ϵ, δ)-query complexity*) Given an optimization algorithm $\mathcal{A}$, an objective function f, and constants $\epsilon > 0$ and $0 < \delta < 1$, the (ϵ, δ)-query complexity of $\mathcal{A}$ on f, denoted as $T_{\mathcal{A},f}(\epsilon, \delta)$, is the minimum number of function evaluations T such that

$$\Pr(f(\mathcal{A}_T) - f(\boldsymbol{x}^*) \leq \epsilon) \geq 1 - \delta, \tag{4.1}$$

where $\mathcal{A}_T$ denotes the solution returned by $\mathcal{A}$ after T function evaluations.

The (ϵ, δ)-query complexity provides a way to quantify the efficiency of an optimization algorithm in terms of the number of function evaluations required to achieve a desired level of accuracy with high probability. It allows us to compare different algorithms and analyze their performance guarantees.

4.3 Performance Bound for SAL Framework

We start by deriving a general upper bound on the (ϵ, δ)-query complexity of the SAL framework. The bound depends on two key quantities: the average success probability of sampling from the learned model and the number of samples required to achieve a certain success probability.

Theorem 4.1 *(General performance bound for SAL) Given an objective function* f *and constants* $\epsilon > 0$ *and* $0 < \delta < 1$, *the* (ϵ, δ)*-query complexity of a SAL algorithm* $\mathcal{A}$ *on* f *is upper bounded by*

$$T_{\mathcal{A},f}(\epsilon, \delta) \leq O\left(\max\left\{\frac{1}{(1-\lambda)\mathrm{Pr}_u + \lambda\overline{\mathrm{Pr}}_h} \ln\frac{1}{\delta},\ m_0 + \sum_{t=1}^{T} m_{\mathrm{Pr}_{h_t}}\right\}\right), \tag{4.2}$$

where λ *is the balancing parameter,*

$$\mathrm{Pr}_u = \int_{D_{\alpha^*}} \mathcal{U}_X(\boldsymbol{x})\,\mathrm{d}\boldsymbol{x} \quad \textit{and} \quad \mathrm{Pr}_{h_t} = \int_{D_{\alpha^*}} \mathcal{T}_{h_t}(\boldsymbol{x})\,\mathrm{d}\boldsymbol{x}$$

are the success probability of uniform sampling and hypothesis distribution, respectively,

$$\overline{\mathrm{Pr}}_h = \frac{\sum_{t=1}^{T} m_t \cdot \mathrm{Pr}_{h_t}}{\sum_{t=1}^{T} m_t}$$

is the average success probability of sampling from the learned hypothesis, $m_{\mathrm{Pr}_{h_t}}$ *is the sample size required to obtain* Pr_{h_t}, *and* $D_{\alpha^*} = \{\boldsymbol{x} \in X \mid f(\boldsymbol{x}) \leq \alpha^*\}$.

Proof The $m_0 + \sum_{t=1}^{T} m_{\mathrm{Pr}_{h_t}}$ part is natural. We prove the rest part of the bound. Let's consider the probability that after T iterations, the SAL algorithm outputs a bad solution $\boldsymbol{x}$ such that $f(\boldsymbol{x}) > \alpha^*$. Since $\boldsymbol{x}$ is the best solution among all sampled examples, the probability is the joint of events that every step of the sampling does not generate such a good solution.

Case 1. For the sampling from the uniform distribution over the whole solution space X, the probability of failure is $1 - \mathrm{Pr}_u$.

Case 2. For the sampling from the learned hypothesis $\mathcal{T}(h_t)$, the probability of failure is denoted as $1 - \mathrm{Pr}_{h_t}$.

Since every sampling is independent, we can expand the probability of overall failures, i.e.,

$$
\begin{aligned}
&\Pr(f(\boldsymbol{x}) > \alpha^*) \\
&= (1 - \Pr_u)^{m_0} \cdot \prod_{t=1}^{T} \sum_{i=0}^{m_t} \binom{m_t}{i} (1-\lambda)^i \lambda^{m_t - i} (1 - \Pr_u)^i (1 - \Pr_{h_t})^{m_t - i} \\
&= (1 - \Pr_u)^{m_0} \prod_{t=1}^{T} \left((1-\lambda)(1-\Pr_u) + \lambda(1 - \Pr_{h_t})\right)^{m_t} \\
&= (1 - \Pr_u)^{m_0} \prod_{t=1}^{T} \left(1 - (1-\lambda)\Pr_u - \lambda \Pr_{h_t}\right)^{m_t} \\
&\le e^{-\Pr_u \cdot m_0} \prod_{t=1}^{T} e^{-((1-\lambda)\Pr_u m_t + \lambda \Pr_{h_t} m_t)} \\
&= e^{-\left(\Pr_u \cdot m_0 + (1-\lambda)\sum_{t=1}^{T} \Pr_u m_t + \lambda \sum_{t=1}^{T} \Pr_{h_t} m_t\right)},
\end{aligned} \tag{4.3}
$$

where the inequality is by $(1 - x) \le e^{-x}$ for $x \in [0, 1]$.

At the same time, letting $\Pr(f(\boldsymbol{x}) > \alpha^*) < \delta$, we get

$$
e^{-\left(\Pr_u \cdot m_0 + (1-\lambda)\sum_{t=1}^{T} \Pr_u m_t + \lambda \sum_{t=1}^{T} \Pr_{h_t} m_t\right)} < \delta, \tag{4.4}
$$

which results in the theorem by noticing that $m_\Sigma > \sum_{t=1}^{T} m_t$. □

Some readers may question the assumption of independence in iterative sampling, given that it is natural for the hypothesis learned in one iteration to be influenced by the samples drawn in previous iterations. This is a valid concern since, in many cases, the results of each iteration are used to refine the sampling process in subsequent steps, potentially introducing dependencies.

However, it is important to clarify that in this context, we are focusing solely on the sampling process itself, not the hypothesis or model that is derived from the samples. The independence we refer to applies to the act of sampling, where each set of samples is considered to be drawn independently from the underlying distribution, regardless of the hypothesis generated in prior iterations. This means that while the hypothesis evolves based on previous iterations, the sampling steps can still be treated as independent actions under the statistical assumptions of the framework.

That said, the calculation of probabilities and any potential dependencies arising from the iterative learning process are more complex. These will be discussed in detail later in the chapter, where we explore how the interplay between sampling-and-learning affects the overall convergence and performance of the algorithm.

4.4 Performance Bound for SAC Framework

To further proceed with the analysis, we consider the SAC framework in this section. According to Theorem 4.1, we need to estimate an upper bound of $\overline{\Pr}_h$, i.e., how likely the distribution $\mathcal{T}_{h_t}$ will lead to a good solution.

We have a lower bound on the success probability as in Lemma 4.1, which implies that without any prior knowledge, uniform distribution is the best worst case.

Lemma 4.1 *For any minimization problem f, any approximation level $\alpha^* > 0$, any hypothesis h, the probability that a sample drawn from an arbitrary distribution $\mathcal{T}_h$ defined on D_h will lead to a solution in D_{α^*} is lower bounded as*

$$\Pr_h = \Pr(\boldsymbol{x} \in D_{\alpha^*} \mid \boldsymbol{x} \sim \mathcal{U}_{D_h}) \geq \frac{\mu(D_{\alpha^*} \cap D_h)}{\mu(D_h)} - \sqrt{2D_{KL}(\mathcal{T}_h \| \mathcal{U}_{D_h})}$$

Proof The proof starts from the definition of the probability,

$$\begin{aligned}
\Pr_h &= \int_{D_h} \mathcal{T}_h(\boldsymbol{x}) \cdot \mathbb{I}(\boldsymbol{x} \in D_{\alpha^*})\,\mathrm{d}\boldsymbol{x} \\
&= \int_{D_h} (\mathcal{T}_h(\boldsymbol{x}) - \mathcal{U}_{D_h}(\boldsymbol{x}) + \mathcal{U}_{D_h}(\boldsymbol{x})) \cdot \mathbb{I}(\boldsymbol{x} \in D_{\alpha^*})\,\mathrm{d}\boldsymbol{x} \\
&= \frac{\mu(D_{\alpha^*} \cap D_h)}{\mu(D_h)} + \int_{D_h} (\mathcal{T}_h(\boldsymbol{x}) - \mathcal{U}_{D_h}(\boldsymbol{x})) \cdot \mathbb{I}(\boldsymbol{x} \in D_{\alpha^*})\,\mathrm{d}\boldsymbol{x} \\
&\geq \frac{\mu(D_{\alpha^*} \cap D_h)}{\mu(D_h)} - \sup_{f:X\to[-1,1]} \int_{D_h} |f(\boldsymbol{x})\mathcal{T}_h(\boldsymbol{x}) - f(\boldsymbol{x})\mathcal{U}_{D_h}(\boldsymbol{x})|\,\mathrm{d}\boldsymbol{x} \\
&\geq \frac{\mu(D_{\alpha^*} \cap D_h)}{\mu(D_h)} - \sqrt{2D_{KL}(\mathcal{T}_h \| \mathcal{U}_{D_h})},
\end{aligned} \tag{4.5}$$

where the last inequality is by Pinsker's inequality. □

We cannot determine D_h, but we know that h is derived by a binary classification learning algorithm from a data set which is labeled according to D_{α_t} for given t. A classifier may not be completely correct. We denote $\epsilon_{\mathcal{D}}$ as the generalization error of h under the distribution $\mathcal{D}$, defined as

$$\epsilon_{\mathcal{D}} = \int_X \mathcal{D}(\boldsymbol{x})\mathbb{I}\Big(h(\boldsymbol{x}) \neq (2\mathbb{I}(\boldsymbol{x} \in D_{\alpha_t}) - 1)\Big)\,\mathrm{d}\boldsymbol{x}. \tag{4.6}$$

Meanwhile, we denote $\hat{\epsilon}_{\mathcal{D}}$ as the training error of h, which is the error on the data set drawn from the distribution $\mathcal{D}$.

For binary classification, we know that the generalization error, which is the expected misclassification rate, can be bounded above by the training error, which is the misclassification rate in the seen examples, as well as the generalization gap involving the complexity of the hypothesis space indicated by the VC-dimension [3], as in Lemma 4.2.

Lemma 4.2 ([3]) *Let $\mathcal{H} = \{h : X \to \{-1, +1\}\}$ be the hypothesis space containing a family of binary classification functions and $VC(\mathcal{H}) = d$. If there exist m samples i.i.d. from X according to some fixed unknown distribution $\mathcal{D}$, then, $\forall\ h \in \mathcal{H}$ and $\forall\ 0 < \eta < 1$, the following upper bound holds true with probability at least $1 - \eta$:*

$$\epsilon_{\mathcal{D}} \leq \hat{\epsilon}_{\mathcal{D}} + \sqrt{8m^{-1}\left(d \log\left(2emd^{-1}\right) + \log\left(4\eta^{-1}\right)\right)} \tag{4.7}$$

where $\epsilon_{\mathcal{D}}$ is the expected error rate of h over $\mathcal{D}$ and $\hat{\epsilon}_{\mathcal{D}}$ is the error rate in the sampled examples from $\mathcal{D}$. When $\hat{\epsilon}_{\mathcal{D}} = 0$,

$$\epsilon_{\mathcal{D}} \leq 2m^{-1}\left(d \log\left(2emd^{-1}\right) + \log\left(2\eta^{-1}\right)\right). \tag{4.8}$$

Again by Pinsker's inequality, we know that the error $\epsilon_{\mathcal{D}}$ under the distribution $\mathcal{D}$ can be converted to the error $\epsilon_{\mathcal{U}}$ under the uniform distribution, as

$$\begin{aligned} \epsilon_{\mathcal{U}} &\leq \epsilon_{\mathcal{D}} + \sqrt{2D_{KL}(\mathcal{D}\|\mathcal{U})} \\ &\leq \hat{\epsilon}_{\mathcal{D}} + \sqrt{\frac{8}{m}\left(d \log 2emd^{-1} + \log 4\eta^{-1}\right)} + \sqrt{2D_{KL}(\mathcal{D}\|\mathcal{U})}, \end{aligned} \tag{4.9}$$

where we only take into account the event that the generalization inequality holds with probability $1 - \eta$. For simplicity, we denote the right-hand side as $\Psi^{m,\eta}_{\hat{\epsilon}_{\mathcal{D}},d,D_{KL}(\mathcal{D}\|\mathcal{U})}$, which decreases with m and η, and increases with $\hat{\epsilon}_{\mathcal{D}}$, d, and $D_{KL}(\mathcal{D}\|\mathcal{U})$.

We can use this result to eliminate the need for D_h in Lemma 4.1. In every iteration of SAC algorithms, there are m_t samples collected.

Theorem 4.2 *For any minimization problem f, any constant $\eta > 0$, and any approximation level $\alpha^* > 0$, the average success probability of sampling from the learned hypothesis of any SAC algorithm is bounded below as*

$$\overline{\Pr}_h \geq \frac{1-\eta}{\sum_{t=1}^{T} m_t} \sum_{t=1}^{T} m_t \left(\frac{\mu(D_{\alpha^*}) - 2\Psi^{m_t,\eta}_{\hat{\epsilon}_{\mathcal{D}_t},d,D_{KL}(\mathcal{D}_t\|\mathcal{U}_X)}}{\mu(D_{\alpha_t}) + \Psi^{m_t,\eta}_{\hat{\epsilon}_{\mathcal{D}_t},d,D_{KL}(\mathcal{D}_t\|\mathcal{U}_X)}} - \sqrt{2D_{KL}(\mathcal{T}_{h_t}\|\mathcal{U}_{D_{h_t}})} \right), \tag{4.10}$$

where $\mathcal{D}_t = \lambda\mathcal{T}_{h_t} + (1-\lambda)\mathcal{U}_X$ is the sampling distribution at iteration t, $\hat{\epsilon}_{\mathcal{D}_t}$ is the training error rate of h_t, and d is the VC-dimension of the learning algorithm.

Proof We follow Lemma 4.1, and examine the terms $\mu(D_{\alpha^*} \cap D_{h_t})$ and $\mu(D_{h_t})$. By set operators,

$$\begin{aligned} \mu(D_{\alpha^*} \cap D_{h_t}) &= \mu(D_{\alpha^*} \cup D_{h_t}) - \mu(D_{\alpha^*} \Delta D_{h_t}) \\ &\geq \mu(D_{\alpha^*} \cup D_{h_t}) - \mu(D_{\alpha^*} \Delta D_{\alpha_t}) - \mu(D_{\alpha_t} \Delta D_{h_t}) \\ &= \mu(D_{\alpha^*} \cup D_{h_t}) - \mu(D_{\alpha^*} \Delta D_{\alpha_t}) - \epsilon_{\mathcal{U},t} \\ &= \mu(D_{\alpha^*} \cup D_{h_t}) + \mu(D_{\alpha^*}) - \mu(D_{\alpha_t}) - \epsilon_{\mathcal{U},t}, \end{aligned} \tag{4.11}$$

where Δ is the symmetric difference operator of two sets, the first inequality is by the triangle inequality, and the last equation is by the fact that D_{α^*} is contained in D_{α_t}.

Since $|\mu(D_{h_t}) - \mu(D_{\alpha_t})| \leq \mu(D_{h_t} \Delta D_{\alpha_t}) = \epsilon_{\mathcal{U},t}$, we can bound $\mu(D_{h_t})$ as $\mu(D_{\alpha_t}) + \epsilon_{\mathcal{U},t} \geq \mu(D_{h_t}) \geq \mu(D_{\alpha_t}) - \epsilon_{\mathcal{U},t}$.

Applying the above bounds to Lemma 4.1, the success probability of sampling from h_t is lower bounded as

$$
\begin{aligned}
\Pr_{h_t} &\geq \frac{\mu(D_{\alpha^*} \cup D_{h_t})}{\mu(D_{h_t})} - \sqrt{2D_{KL}(\mathcal{T}_h \| \mathcal{U}_{D_h})} \\
&\geq \frac{\mu(D_{\alpha^*} \cup D_{h_t}) + \mu(D_{\alpha^*}) - \mu(D_{\alpha_t}) - \epsilon_{\mathcal{U},t}}{\mu(D_{\alpha_t}) + \epsilon_{\mathcal{U},t}} - \sqrt{2D_{KL}(\mathcal{T}_{h_t} \| \mathcal{U}_{D_{h_t}})} \\
&\geq \frac{\mu(D_{h_t}) + \mu(D_{\alpha^*}) - \mu(D_{\alpha_t}) - \epsilon_{\mathcal{U},t}}{\mu(D_{\alpha_t}) + \epsilon_{\mathcal{U},t}} - \sqrt{2D_{KL}(\mathcal{T}_{h_t} \| \mathcal{U}_{D_{h_t}})} \\
&\geq \frac{\mu(D_{\alpha^*}) - 2\epsilon_{\mathcal{U},t}}{\mu(D_{\alpha_t}) + \epsilon_{\mathcal{U},t}} - \sqrt{2D_{KL}(\mathcal{T}_{h_t} \| \mathcal{U}_{D_{h_t}})}. \qquad (4.12)
\end{aligned}
$$

Substituting this lower bound and the probability $1 - \eta$ of the generalization bound into $\overline{\Pr}_h$ yields the theorem. □

Combining Theorems 4.1 and 4.2 results in an upper bound on the sampling complexity of SAC algorithms. Although the expression is sophisticated, it can still reveal relative variables that generally affect the complexity. One could design various distributions for $\mathcal{T}_h$ to sample potential solutions; however, without any a priori knowledge, uniform sampling will lead to the best worst-case performance. Meanwhile, without any a priori knowledge, a small training error at each stage from a learning algorithm with a small VC-dimension will improve the performance.

4.5 Error-Target Dependence and Shrinking Rate

Next, we focus on the SAC framework and derive a more specific upper bound on its (ϵ, δ)-query complexity. The bound reveals two critical factors that influence the performance of the SAC framework: the error-target dependence and the shrinking rate.

Definition 4.3 (*Error-target θ-dependence*) Given a SAC algorithm $\mathcal{A}$, the error-target dependence of $\mathcal{A}$ is the smallest constant $\theta \geq 0$ such that for all $\epsilon > 0$ and all iterations t,

$$
\begin{aligned}
&\mu(D_\epsilon) \cdot \mu(D_{\alpha_t} \Delta D_{h_t}) - \theta\mu(D_\epsilon) \\
&\qquad \leq \mu(D_\epsilon \cap (D_{\alpha_t} \Delta D_{h_t})) \\
&\qquad\qquad \leq \mu(D_\epsilon) \cdot \mu(D_{\alpha_t} \Delta D_{h_t}) + \theta\mu(D_\epsilon), \qquad (4.13)
\end{aligned}
$$

where the operator Δ is the symmetric difference of two sets, i.e., $A_1 \Delta A_2 = (A_1 \cup A_2) - (A_1 \cap A_2)$. It characterizes, when sampling a solution x from $\mathcal{U}_X$, the dependence between the random variable that whether $\boldsymbol{x} \in D_{\alpha_t} \Delta D_{h_t}$ and the random variable that whether $\boldsymbol{x} \in D_\epsilon$.

The error-target dependence measures the alignment between the classification error and the target ϵ-optimal set. A smaller θ indicates a better alignment, with $\theta = 0$ implying that the classification error is independent of the target set.

Definition 4.4 (γ-*Shrinking rate*) Given a SAC algorithm $\mathcal{A}$, the shrinking rate of $\mathcal{A}$ is the largest constant $\gamma > 0$ such that for all iterations t,

$$\mu(D_{h_t}) \leq \gamma \mu(D_{\alpha_t}). \tag{4.14}$$

The shrinking rate quantifies the reduction in the volume of the positive region of the learned hypothesis compared to the α_t-sublevel set. Note that the rate does not imply any overlap between the positive region and the α_t-sublevel set.

Using these definitions, we can derive a performance bound for the SAC framework with the parameters θ and γ. Note these parameters do not weaken the generality of the analysis.

Theorem 4.3 *(Parameterized bound for SAC) Given an objective function f and constants $\epsilon > 0$ and $0 < \delta < 1$, the (ϵ, δ)-query complexity of a SAC algorithm $\mathcal{A}$ on f with error-target dependence θ and shrinking rate γ is upper bounded by*

$$T_{\mathcal{A},f}(\epsilon, \delta) \leq O\left(\frac{1}{\mu(D_\epsilon)}\left((1-\lambda) + \frac{\lambda}{\gamma T}\sum_{t=1}^{T}\frac{1-\theta-\frac{\epsilon_{\mathcal{D}_t}}{1-\lambda}}{\mu(D_{\alpha_t})}\right)^{-1}\ln\frac{1}{\delta}\right), \tag{4.15}$$

where $\lambda \in [0, 1]$ is the balancing parameter and ϵ_t is an upper bound on the generalization error of the learned hypothesis h_t at iteration t.

Theorem 4.3 discloses that the error-target θ-dependence and the γ-shrinking rate are two important factors. It can be observed that the smaller θ and γ, the better the query complexity.

The proof of Theorem 4.3 follows a similar structure to the proof of Theorem 4.1, with additional steps to incorporate the error-target dependence and the shrinking rate. To prove this theorem, our strategy is to refine the bound of $\mu(D_\epsilon \cap D_{h_t})$ under the *error-target θ-dependence* condition and the bound of $\mu(D_{h_t})$ under the *γ-shrinking rate* condition, respectively.

Lemma 4.3 *For the classifier-based optimization algorithms under the condition of error-target θ-dependence,*

$$\mu(D_\epsilon \cap D_{h_t}) \geq \mu(D_\epsilon) \cdot (1 - \epsilon_{\mathcal{U}_X,t} - \theta)$$

holds for all t, where $\epsilon_{\mathcal{U}_X,t}$ is the generalization error of h_t under $\mathcal{U}_X$ in iteration t.

Proof Assume w.l.o.g. that $\epsilon \le \alpha_t$ for all t, we have

$$\begin{aligned}\mu(D_\epsilon \cap D_{h_t}) &= \mu(D_\epsilon) - \mu(D_\epsilon \cap (D_{\alpha_t} \Delta D_{h_t}))\\ &\ge \mu(D_\epsilon) - \mu(D_\epsilon) \cdot \mu(D_{\alpha_t} \Delta D_{h_t}) - \theta\mu(D_\epsilon)\\ &= \mu(D_\epsilon)(1 - \mu(D_{\alpha_t} \Delta D_{h_t}) - \theta),\end{aligned}$$

where the first equality is by $D_\epsilon \subseteq D_{\alpha_t}$, and the first inequality is by the condition of *error-target θ-dependence*.

Let $\epsilon_{\mathcal{U}_X,t}$ denote the generalization error of h_t under $\mathcal{U}_X$ in iteration t, it can be verified directly that $\epsilon_{\mathcal{U}_X,t} = \mu(D_{\alpha_t} \Delta D_{h_t})$ under 0–1 loss. Thus, we have $\mu(D_\epsilon \cap D_{h_t}) \ge \mu(D_\epsilon)(1 - \epsilon_{\mathcal{U}_X,t} - \theta)$. □

In order to refine Lemma 4.3, i.e., lower bound $\mu(D_\epsilon \cap D_{h_t})$ using the generalization error of h_t under the true sampling distribution $\mathcal{D}_t = \lambda\mathcal{U}_{D_{h_t}} + (1-\lambda)\mathcal{U}_X$ instead of $\mathcal{U}_X$, we need Lemma 4.4 below. It gives a relationship between $\epsilon_{\mathcal{U}_X,t}$ and $\epsilon_{\mathcal{D}_t}$, where $\epsilon_{\mathcal{D}_t}$ is the generalization error of h_t under $\mathcal{D}_t$ in iteration t.

Lemma 4.4 *For any $h_t \in \mathcal{H}$, let $\mathcal{D}_t = \lambda\mathcal{U}_{D_{h_t}} + (1-\lambda)\mathcal{U}_X$, it holds for all t that $\epsilon_{\mathcal{U}_X,t} \le \epsilon_{\mathcal{D}_t}/(1-\lambda)$, where $\lambda \in (0, 1)$.*

Proof We only consider continuous domains situation and omit finite discrete domains situation since the proof procedure is quite similar. Let $D_{\neq,t}$ be the region where h_t makes mistakes. Splitting $D_{\neq,t}$ into $D^+_{\neq,t} = D_{\neq,t} \cap D_{h_t}$ and $D^-_{\neq,t} = D_{\neq,t} - D^+_{\neq,t}$, we can calculate the probability density $\mathcal{D}_t(x) = \lambda\frac{1}{\mu(D_{h_t})} + (1-\lambda)\frac{\mu(D_{h_t})}{\mu(X)}\frac{1}{\mu(D_{h_t})} = \lambda\frac{1}{\mu(D_{h_t})} + (1-\lambda)\frac{1}{\mu(X)}$ for any $x \in D^+_{\neq,t}$, and $\mathcal{D}_t(x) = (1-\lambda)\frac{\mu(X-D_{h_t})}{\mu(X)}\frac{1}{\mu(X-D_{h_t})} = (1-\lambda)\frac{1}{\mu(X)}$ for any $x \in D^-_{\neq,t}$. Thus,

$$\begin{aligned}\epsilon_{\mathcal{D}_t} &= \int_X \mathcal{D}_t(\boldsymbol{x}) \cdot \mathbb{I}[h_t \text{ makes mistake on } \boldsymbol{x}]\,\mathrm{d}\boldsymbol{x}\\ &= \int_{D_{\neq,t}} \mathcal{D}_t(\boldsymbol{x})\,\mathrm{d}\boldsymbol{x} = \int_{D^+_{\neq,t}} \mathcal{D}_t(\boldsymbol{x})\,\mathrm{d}\boldsymbol{x} + \int_{D^-_{\neq,t}} \mathcal{D}_t(\boldsymbol{x})\,\mathrm{d}\boldsymbol{x}\\ &\ge \int_{D^+_{\neq,t}} (1-\lambda)\frac{1}{\mu(X)}\,\mathrm{d}\boldsymbol{x} + \int_{D^-_{\neq,t}} (1-\lambda)\frac{1}{\mu(X)}\,\mathrm{d}\boldsymbol{x}\\ &= (1-\lambda)\epsilon_{\mathcal{U}_X,t},\end{aligned}$$

which proves the lemma. □

Combining Lemma 4.4 with Lemma 4.3, we can conclude that $\mu(D_\epsilon \cap D_{h_t}) \ge \mu(D_\epsilon) \cdot (1 - \theta - \epsilon_{\mathcal{D}_t}/(1-\lambda))$. Meanwhile, the *$\gamma$-shrinking rate* condition admits $\mu(D_{h_t}) \le \gamma\mu(D_{\alpha_t})$ for all t directly. So far, the proof of Theorem 4.3 becomes clear, and is presented as follows.

Proof (**Theorem** 4.3) By Lemma 4.1 and the assumption of $\mathcal{T}_{h_t} = \mathcal{U}_{D_{h_t}}$, $D_{KL}(\mathcal{T}_{h_t} \| \mathcal{U}_{D_{h_t}}) = 0$ and thus $\Pr_{h_t} \geq \mu(D_\epsilon \cap D_{h_t})/\mu(D_{h_t})$ for all t. Combining it with the refined bounds of $\mu(D_\epsilon \cap D_{h_t})$ and $\mu(D_{h_t})$ results in that

$$\Pr_{h_t} \geq \frac{(1 - \theta - \epsilon_{\mathcal{D}_t}/(1 - \lambda)) \cdot \mu(D_\epsilon)}{\gamma \cdot \mu(D_{\alpha_t})}.$$

Finally, by the definition of $\overline{\Pr}_h$ and Theorem 4.1 we prove the theorem. □

Theorem 4.3 highlights the importance of the error-target dependence and the shrinking rate in determining the performance of the SAC framework. A smaller error-target dependence (θ) and a smaller shrinking rate (γ) lead to a tighter upper bound on the (ϵ, δ)-query complexity, indicating better sample efficiency.

Despite the generality of the performance bound, we still need to understand on what kind of optimization problems SAC algorithms can perform well. We investigate two classes of objective functions: functions satisfying the local Lipschitz condition and functions with bounded packing and covering numbers. We will show in the following sections that SAC algorithms can be efficient on the two classes.

4.6 Functions with Local Lipschitz Continuity

We find that a class of functions $\mathcal{F}_L \subseteq \mathcal{F}$ satisfying the *local Lipschitz* continuity (Definition 4.5) can be efficiently optimized by SAC algorithms with error-target dependence $\theta < 1$ and shrinking rate $\gamma > 0$. For finite discrete domains, we consider $X = \{0, 1\}^n$ and let $\|\boldsymbol{x} - \boldsymbol{y}\|_H$ denote the Hamming distance between $\boldsymbol{x}, \boldsymbol{y} \in \{0, 1\}^n$.

Definition 4.5 (*Local Lipschitz*) Given $f \in \mathcal{F}$, let $\boldsymbol{x}^*$ be a global minimum of f, for all $\boldsymbol{x} \in X$, if $X = \{0, 1\}^n$, then there exist positive constants $\beta_1, \beta_2, L_1, L_2$ such that

$$L_2 \|\boldsymbol{x} - \boldsymbol{x}^*\|_H^{\beta_2} \leq f(\boldsymbol{x}) - f(\boldsymbol{x}^*) \leq L_1 \|\boldsymbol{x} - \boldsymbol{x}^*\|_H^{\beta_1};$$

if X is a compact continuous domains, then there exist positive constants $\beta_1, \beta_2, L_1, L_2$ such that

$$L_2 \|\boldsymbol{x} - \boldsymbol{x}^*\|_2^{\beta_2} \leq f(\boldsymbol{x}) - f(\boldsymbol{x}^*) \leq L_1 \|\boldsymbol{x} - \boldsymbol{x}^*\|_2^{\beta_1}.$$

Let $\mathcal{F}_L^{\beta_1, L_1, \beta_2, L_2} (\subseteq \mathcal{F})$ denote the function class that satisfies the condition.

This condition guarantees that f has a bounded change range around the global minimum $\boldsymbol{x}^*$. Within this constraint, the landscape of f can be quite complex, such as having many local minima.

Note that we can have classification algorithms with the convergence rate of the generalization error $\widetilde{O}(\frac{1}{m})$ ignoring other variables and logarithmic terms [3, 4], where m is the sample size for the learning. Thus we assume that classification algorithms with convergence rate $\widetilde{\Theta}(\frac{1}{m})$ will be employed. We then prove that SAC algorithms have polynomial (ϵ, δ)-query complexity for local Lipschitz problems in both discrete and continuous domains.

Corollary 4.1 *In finite discrete domains $X = \{0, 1\}^n$, given $f \in \mathcal{F}_L^{\beta_1, L_1, \beta_2, L_2}$, $0 < \delta < 1$ and $0 < \epsilon \leq L_1(\frac{n}{2})^{\beta_1}$, for a classification-based optimization algorithm using a classification algorithm with convergence rate $\widetilde{\Theta}(\frac{1}{m})$, under the conditions that error-target dependence $\theta < 1$ and shrinking rate $\gamma > 0$, its (ϵ,δ)-query complexity belongs to $\texttt{poly}(\frac{1}{\epsilon}, n, \frac{1}{\beta_1}, \beta_2, \ln L_1, \ln \frac{1}{L_2}) \cdot \ln \frac{1}{\delta}$.*

Proof Following the proof procedure of Theorem 4.3, letting $\lambda = 1/2$, we have

$$\overline{\Pr}_h \geq \frac{1}{T} \sum_{t=1}^{T} (K_t \cdot \mu(D_\epsilon))/(\gamma \cdot \mu(D_{\alpha_t})),$$

where $K_t = 1 - \theta - 2\epsilon_{\mathcal{D}_t}$. Assume that $\theta < 1$, there must exist a constant $K > 0$ such that $K_t \geq K$ as long as $\epsilon_{\mathcal{D}_t} < (1 - \theta)/2$ for all t.

Under the assumption of employing classification algorithms with convergence rate $\widetilde{\Theta}(\frac{1}{m})$, $\epsilon_{\mathcal{D}_t} < (1 - \theta)/2$ can be guaranteed if the sampled solution size m in each iteration belongs to $\texttt{poly}(\frac{1}{\epsilon}, n)$ [3]. Letting $K' = K/\gamma$, we therefore obtain that

$$\overline{\Pr}_h \geq \frac{1}{T} \sum_{t=1}^{T} \frac{K \cdot \mu(D_\epsilon)}{\gamma \cdot \mu(D_{\alpha_t})} = \frac{K'}{T} \sum_{t=1}^{T} \frac{\mu(D_\epsilon)}{\mu(D_{\alpha_t})}.$$

Since $f \in \mathcal{F}_L^{\beta_1, L_1, \beta_2, L_2}$ has local Lipschitz continuity, we know

$$L_2 \|\boldsymbol{x} - \boldsymbol{x}^*\|_H^{\beta_2} \leq f(\boldsymbol{x}) - f(\boldsymbol{x}^*) \leq L_1 \|\boldsymbol{x} - \boldsymbol{x}^*\|_H^{\beta_1}.$$

Denote $\widetilde{D}_\epsilon = \{\boldsymbol{x} \in X \mid \|\boldsymbol{x} - \boldsymbol{x}^*\|_H^{\beta_1} \leq \frac{\epsilon}{L_1}\}$. It can be verified directly that $\widetilde{D}_\epsilon \subseteq D_\epsilon$ and thus $\mu(\widetilde{D}_\epsilon) \leq \mu(D_\epsilon)$.

Let $\alpha'_t = \alpha_t - f(\boldsymbol{x}^*)$ and we assume that $\alpha'_t > 0$,

$$D_{\alpha_t} = \{\boldsymbol{x} \in X \mid f(\boldsymbol{x}) \leq \alpha_t\} = \{\boldsymbol{x} \in X \mid f(\boldsymbol{x}) - f(\boldsymbol{x}^*) \leq \alpha'_t\}.$$

Denote $\widetilde{D}_{\alpha_t} = \{\boldsymbol{x} \in X \mid \|\boldsymbol{x} - \boldsymbol{x}^*\|_H^{\beta_2} \leq \frac{\alpha'_t}{L_2}\}$. Similarly, we have $D_{\alpha_t} \subseteq \widetilde{D}_{\alpha_t}$ and thus $\mu(D_{\alpha_t}) \leq \mu(\widetilde{D}_{\alpha_t})$.

For simplicity, we assume that $(\frac{\epsilon}{L_1})^{\frac{1}{\beta_1}}$ and $(\frac{\alpha'_t}{L_2})^{\frac{1}{\beta_2}}$ are both positive integers. By the definition of Hamming distance, we have

$$\mu(\widetilde{D}_\epsilon) = \sum_{i=0}^{(\frac{\epsilon}{L_1})^{\frac{1}{\beta_1}}} \binom{n}{i} \quad \text{and} \quad \mu(\widetilde{D}_{\alpha_t}) = \sum_{i=0}^{(\frac{\alpha'_t}{L_2})^{\frac{1}{\beta_2}}} \binom{n}{i}.$$

Let $H(p) = -p \log p - (1-p)\log(1-p)$ which is the binary entropy function of p, where $0 \le p \le 1$ and $H(p) = 0$ for $p = 0, 1$. Then, the following inequality [1] holds for all integers $0 \le k \le n$ with $p = k/n \le 1/2$

$$\frac{1}{1+\sqrt{8np(1-p)}} \cdot 2^{nH(p)} \le \sum_{i=0}^{k} \binom{n}{i} \le 2^{nH(p)}.$$

Since $0 < \epsilon \le L_1(\frac{n}{2})^{\beta_1}$, we have $(\frac{\epsilon}{L_1})^{\frac{1}{\beta_1}} \le \frac{n}{2}$. Meanwhile, choosing $\alpha'_t = \frac{2L_2}{2^t}$ for all t can guarantee that $(\frac{\alpha'_t}{L_2})^{\frac{1}{\beta_2}} \le \frac{n}{2}$ for all t because $(\frac{\alpha'_1}{L_2}) = 1 \le (\frac{n}{2})^{\beta_2}$ for $n \ge 2$. If $n = 1$, we can still choose smaller α'_t s.t. $(\frac{\alpha'_t}{L_2})^{\frac{1}{\beta_2}} \le \frac{n}{2}$, and we omit the details since it is easy to verify. Combing the above statement with the inequality $\overline{\Pr}_h \ge \frac{K'}{T}\sum_{t=1}^{T} \mu(D_\epsilon)/\mu(D_{\alpha_t})$, we have

$$\begin{aligned}
\overline{\Pr}_h &\ge \frac{K'}{T}\sum_{t=1}^{T} \frac{\mu(\widetilde{D}_\epsilon)}{\mu(\widetilde{D}_{\alpha_t})} = \frac{K'}{T}\sum_{t=1}^{T} \frac{\mu(\widetilde{D}_\epsilon)}{\mu(\widetilde{D}_{\alpha_t})} \\
&= \frac{K'}{T}\sum_{t=1}^{T} \frac{\sum_{i=0}^{(\frac{\epsilon}{L_1})^{\frac{1}{\beta_1}}} \binom{n}{i}}{\sum_{i=0}^{(\frac{\alpha'_t}{L_2})^{\frac{1}{\beta_2}}} \binom{n}{i}} \\
&\ge \frac{K'}{T} \cdot \frac{2^{nH\left((\frac{\epsilon}{L_1})^{\frac{1}{\beta_1}}\right)}}{1+\sqrt{8(\frac{\epsilon}{L_1})^{\frac{1}{\beta_1}}(1-(\frac{\epsilon}{L_1})^{\frac{1}{\beta_1}}/n)}} \sum_{t=1}^{T} 2^{-nH\left((\frac{\alpha'_t}{L_2})^{\frac{1}{\beta_2}}\right)}.
\end{aligned}$$

Let the number of iterations T to approach $(\frac{\alpha'_T}{L_2})^{\frac{1}{\beta_2}} = (\frac{\epsilon}{L_1})^{\frac{1}{\beta_1}}$. Solving this equation results in that

$$T = \frac{\beta_2}{\beta_1} \log \frac{L_1}{\epsilon} + 1 \in \mathtt{poly}\left(\frac{1}{\epsilon}, n, \frac{1}{\beta_1}, \beta_2, \log L_1\right).$$

For simplicity, we assume that $\frac{\beta_2}{\beta_1} \log \frac{L_1}{\epsilon} + 1$ is a positive integer and let the SAC algorithm run $T = \frac{\beta_2}{\beta_1} \log \frac{L_1}{\epsilon} + 1$ number of iterations. Now, we can conclude that

$$\overline{\Pr}_h \ge \left(\mathtt{poly}(\frac{1}{\epsilon}, n, \frac{1}{\beta_1}, \beta_2, \log L_1, \log \frac{1}{L_2})\right)^{-1}.$$

Substituting this bound into Theorem 4.1, we have

$$(m+1)T \in \mathtt{poly}(\frac{1}{\epsilon}, n, \frac{1}{\beta_1}, \beta_2, \ln L_1, \ln \frac{1}{L_2}) \cdot \ln \frac{1}{\delta},$$

with probability at least $1-\delta$. Finally, combining the fact that $\epsilon_{\mathcal{D}_t} < (1-\theta)/2$ can be guaranteed with $\texttt{poly}(\frac{1}{\epsilon}, n)$ sampled solutions in each iteration and $T \in \texttt{poly}(\frac{1}{\epsilon}, n, \frac{1}{\beta_1}, \beta_2, \ln L_1)$, the (ϵ,δ)-query complexity of the SAC algorithms belongs to $\texttt{poly}(\frac{1}{\epsilon}, n, \frac{1}{\beta_1}, \beta_2, \ln L_1, \ln \frac{1}{L_2}) \cdot \ln \frac{1}{\delta}$. □

Corollary 4.2 *In compact continuous domains X, given $f \in \mathcal{F}_L^{\beta_1, L_1, \beta_2, L_2}$, $0 < \delta < 1$ and $\epsilon > 0$, for a classification-based optimization algorithm using a classification algorithm with convergence rate $\widetilde{\Theta}(\frac{1}{m})$, under the conditions that error-target dependence $\theta < 1$ and shrinking rate $\gamma > 0$, its (ϵ,δ)-query complexity belongs to $\texttt{poly}(\frac{1}{\epsilon}, n, \frac{1}{\beta_1}, \beta_2, \ln L_1, \ln \frac{1}{L_2}) \cdot \ln \frac{1}{\delta}$.*

The proof is very similar with that of Corollary 4.1, only except that $\mu(\widetilde{D}_\epsilon)$ is the volume of ℓ_2 ball of radius $(\frac{\epsilon}{L_1})^{\frac{1}{\beta_1}}$ in $\mathbb{R}^n$ which is proportional to $(\frac{\epsilon}{L_1})^{\frac{n}{\beta_1}}$, and $\mu(\widetilde{D}_{\alpha_t})$ is the volume of ℓ_2 ball of radius $(\frac{\alpha_t'}{L_2})^{\frac{1}{\beta_2}}$ in $\mathbb{R}^n$ which is proportional to $(\frac{\alpha_t'}{L_2})^{\frac{n}{\beta_2}}$.

4.7 Functions with Bounded Packing and Covering Numbers

More generally, instead of the *local Lipschitz* continuity for compact continuous domains, we present another sufficient condition under which f can be efficiently optimized by classification-based optimization algorithms, using the η-Packing Number and η-Covering Number (Definition 4.6). Recall that $D_\epsilon = \{\boldsymbol{x} \in X \mid f(\boldsymbol{x}) - f(\boldsymbol{x}^*) \le \epsilon\}$ for any $\epsilon > 0$. Let $\alpha_t' = \alpha_t - f(\boldsymbol{x}^*)$ and we assume that $\alpha_t' > 0$.

Definition 4.6 (*η-Packing Number & η-Covering Number*) η-Packing Number is the largest $\mathcal{N}_p \ge 0$ such that, there exists $C_1 > 0$, for all $\epsilon > 0$, the maximal number of disjoint ℓ_2-balls of radius $\eta\epsilon$ contained in D_ϵ with center in D_ϵ is not less than $C_1\epsilon^{-\mathcal{N}_p}$.

Meanwhile, η-Covering Number is the smallest $\mathcal{N}_c \ge 0$ such that, there exists $C_2 > 0$, for all $\epsilon > 0$, the minimal number of ℓ_2-balls of radius $\eta\epsilon$ with center in X covering D_ϵ is not larger than $C_2\epsilon^{-\mathcal{N}_c}$.

Corollary 4.3 *In compact continuous domains X, given $f \in \mathcal{F}$ satisfying $\sum_{t=1}^{T} (\alpha_t')^{\mathcal{N}_c - n} \in \Omega\left(\epsilon^{\mathcal{N}_p - n}\right)$, where $\mathcal{N}_p$ and $\mathcal{N}_c$ are its η-Packing and η-Covering numbers, respectively, $0 < \delta < 1$ and $\epsilon > 0$, for a classification-based optimization algorithm using the classification algorithms with convergence rate $\widetilde{\Theta}(\frac{1}{m})$, under the conditions that error-target dependence $\theta < 1$ and shrinking rate $\gamma > 0$, its (ϵ,δ)-query complexity belongs to $\texttt{poly}(\frac{1}{\epsilon}, n) \cdot \ln \frac{1}{\delta}$.*

Proof By the proof procedure of Theorem 4.3, letting $\lambda = 1/2$, we have $\overline{\Pr}_h \ge \frac{1}{T}\sum_{t=1}^{T}(K_t \cdot \mu(D_\epsilon))/(\gamma \cdot \mu(D_{\alpha_t}))$, where $K_t = 1 - 2\epsilon_{\mathcal{D}_t} - \theta$. Assume that $\theta < 1$, since $K_t = 1 - 2\epsilon_{\mathcal{D}_t} - \theta$ for all t, there must exist a constant $K > 0$ such that $K_t \ge$

K as long as $\epsilon_{\mathcal{D}_t} < (1-\theta)/2$ for all t. Under the assumption of classifier-based optimization using the classification algorithms with convergence rate $\widetilde{\Theta}(\frac{1}{m})$, $\epsilon_{\mathcal{D}_t} < (1-\theta)/2$ can be guaranteed if the sampled solution size m in each iteration belongs to $\texttt{poly}(\frac{1}{\epsilon}, n)$ [3]. Letting $K' = K/\gamma$, we therefore obtain that $\overline{\Pr}_h \geq \frac{1}{T}\sum_{t=1}^{T}(K \cdot \mu(D_\epsilon))/(\gamma \cdot \mu(D_{\alpha_t})) = \frac{K'}{T}\sum_{t=1}^{T}\mu(D_\epsilon)/\mu(D_{\alpha_t})$.

Recall that $D_\epsilon = \{\boldsymbol{x} \in X \mid f(\boldsymbol{x}) - f(\boldsymbol{x}^*) \leq \epsilon\}$ for any $\epsilon > 0$. Let $\alpha_t' = \alpha_t - f(\boldsymbol{x}^*)$ and we assume that $\alpha_t' > 0$, thus, $D_{\alpha_t} = \{\boldsymbol{x} \in X \mid f(\boldsymbol{x}) \leq \alpha_t\} = \{\boldsymbol{x} \in X \mid f(\boldsymbol{x}) - f(\boldsymbol{x}^*) \leq \alpha_t'\}$. Let $V(D_\epsilon)$, $V(D_{\alpha_t})$ and $V(\eta\epsilon)$ denote the volume of D_ϵ, D_{α_t} and ℓ_2 ball of radius $\eta\epsilon$ in $\mathbb{R}^n$ respectively. By the definition of $\mathcal{N}_p$ and $\mathcal{N}_c$, we have

$$C_1\epsilon^{-\mathcal{N}_p} \cdot V(\eta\epsilon) \leq V(D_\epsilon) = \mu(D_\epsilon) \leq C_2\epsilon^{-\mathcal{N}_c} \cdot V(\eta\epsilon),$$

$$C_1(\alpha_t')^{-\mathcal{N}_p} \cdot V(\eta\alpha_t') \leq V(D_{\alpha_t}) = \mu(D_{\alpha_t}) \leq C_2(\alpha_t')^{-\mathcal{N}_c} \cdot V(\eta\alpha_t').$$

Note that the volume of ℓ_2 ball of radius $\eta\epsilon$ in $\mathbb{R}^n$ is $\frac{\pi^{n/2}}{\Gamma(n/2+1)}(\eta\epsilon)^n$. Combing it with the inequality $\overline{\Pr}_h \geq \frac{K'}{T}\sum_{t=1}^{T}\mu(D_\epsilon)/\mu(D_{\alpha_t})$, we have

$$\begin{aligned}
\overline{\Pr}_h &\geq \frac{K'}{T}\sum_{t=1}^{T}\frac{\mu(D_\epsilon)}{\mu(D_{\alpha_t})} = \frac{K'}{T}\sum_{t=1}^{T}\frac{\mu(D_\epsilon)}{\mu(D_{\alpha_t})} \\
&\geq \frac{K'}{T}\sum_{t=1}^{T}\frac{C_1\epsilon^{-\mathcal{N}_p} \cdot V(\eta\epsilon)}{C_2(\alpha_t')^{-\mathcal{N}_c} \cdot V(\eta\alpha_t')} = \frac{K'}{T}\sum_{t=1}^{T}\frac{C_1\epsilon^{-\mathcal{N}_p} \cdot (\eta\epsilon)^n}{C_2(\alpha_t')^{-\mathcal{N}_c} \cdot (\eta\alpha_t')^n} \\
&= \frac{C_1K'}{C_2T}\sum_{t=1}^{T}\frac{\epsilon^{n-\mathcal{N}_p}}{(\alpha_t')^{n-\mathcal{N}_c}} = \frac{C_1K' \cdot \epsilon^{n-\mathcal{N}_p}}{C_2T}\sum_{t=1}^{T}(\alpha_t')^{\mathcal{N}_c-n}.
\end{aligned}$$

Let $T \in \texttt{poly}(\frac{1}{\epsilon}, n)$, if the problem $f \in \mathcal{F}$ satisfying $\sum_{t=1}^{T}(\alpha_t')^{\mathcal{N}_c-n} \in \Omega\left(\epsilon^{\mathcal{N}_p-n}\right)$, we can conclude that $\overline{\Pr}_h \geq \left(\texttt{poly}(\frac{1}{\epsilon}, n)\right)^{-1}$.

Substituting $\overline{\Pr}_h \geq \left(\texttt{poly}(\frac{1}{\epsilon}, n)\right)^{-1}$ into Theorem 4.1, we have $(m+1)T \in \texttt{poly}(\frac{1}{\epsilon}, n) \cdot \ln\frac{1}{\delta}$, with probability at least $1-\delta$. Finally, combining the fact that $\epsilon_{\mathcal{D}_t} < (1-\theta)/2$ can be guaranteed with $\texttt{poly}(\frac{1}{\epsilon}, n)$ sampled solutions in each iteration and $T \in \texttt{poly}(\frac{1}{\epsilon}, n)$, the (ϵ,δ)-query complexity of the classifier-based optimization algorithms belongs to $\texttt{poly}(\frac{1}{\epsilon}, n) \cdot \ln\frac{1}{\delta}$. □

For continuous domains, we have $\mathcal{N}_p \leq \mathcal{N}_c$. Because if we let $V(D_\epsilon)$ and $V(\eta\epsilon)$ denote the volume of D_ϵ and ℓ_2 ball of radius $\eta\epsilon$ in $\mathbb{R}^n$ respectively, then it holds that $C_1\epsilon^{-\mathcal{N}_p} \cdot V(\eta\epsilon) \leq V(D_\epsilon) \leq C_2\epsilon^{-\mathcal{N}_c} \cdot V(\eta\epsilon)$. It is worthwhile to point out that if $\mathcal{N}_c = \mathcal{N}_p = n$, the condition $\sum_{t=1}^{T}(\alpha_t')^{\mathcal{N}_c-n} \in \Omega\left(\epsilon^{\mathcal{N}_p-n}\right)$ can always be satisfied, which implies that classification-based optimization is efficient on this class of functions.

4.8 Summary

In this chapter, we presented a theoretical analysis of the sampling-and-learning (SAL) and sampling-and-classification (SAC) frameworks for derivative-free optimization. We introduced the concept of (ϵ, δ)-query complexity to measure the sample efficiency of these frameworks and derived general performance bounds in terms of the success probability of sampling from the learned models.

For the SAC framework, we identified two key factors that influence its performance: the error-target dependence and the shrinking rate. We showed that under certain conditions on these factors, the SAC framework can achieve a polynomial query complexity for functions satisfying the local Lipschitz condition and functions with bounded packing and covering numbers.

Our theoretical results provide insights into the behavior of the SAL and SAC frameworks and highlight the importance of the alignment between the learned models and the target ϵ-optimal set. They also shed light on the role of the problem geometry and the complexity of the objective function in determining the sample efficiency of these frameworks.

The analysis presented in this chapter has several practical implications. It provides guidance on designing effective learning algorithms for the SAL and SAC frameworks by focusing on reducing the generalization error and improving the alignment with the target set. It also suggests strategies for selecting the thresholds and the balancing parameter to achieve a desired trade-off between exploration and exploitation.

Furthermore, our theoretical results can help develop new derivative-free optimization algorithms by incorporating the insights gained from the analysis. For example, one could design adaptive strategies for setting the thresholds based on the estimated error-target dependence or the shrinking rate. One could also explore hybrid approaches that combine the strengths of different learning algorithms or incorporate prior knowledge about the problem geometry. We have also noticed the latest study proposed the *hypothesis-target η-shattering rate* to replace the *error-target θ-dependence* for achieving tighter bounds [2].

References

1. Ash RB (1990) Information theory. Dover Publications Inc., New York
2. Han T, Li J, Guo Z, Jin Y (2025) Scalable acceleration for classification-based derivative-free optimization. In: Proceedings of the 39st AAAI conference on artificial intelligence (AAAI'25), Philadelphia, PA
3. Kearns MJ, Vazirani UV (1994) An introduction to computational learning theory. MIT Press, Cambridge, MA
4. Vapnik V (2000) The nature of statistical learning theory. Springer Science & Business Media

Chapter 5
Basic Algorithm

Abstract This chapter introduces the Racos (RAndomized COordinate Shrinking) optimization algorithm, a novel approach designed to address complex optimization problems in both continuous and discrete search spaces. Building on the theoretical insights from the previous chapter, Racos minimizes critical factors such as error-target dependence and shrinking rate to enhance optimization efficiency. The algorithm integrates a randomized coordinate shrinking classification technique, which effectively balances exploration and exploitation in the search process. The chapter is structured as follows: Sect. 5.1 details the Racos algorithm, Sect. 5.2 presents empirical evaluations on benchmark functions, Sect. 5.3 applies Racos to spectral clustering tasks, and Sect. 5.4 examines its performance in classification tasks using Ramp loss. Experimental results demonstrate Racos's superiority over state-of-the-art derivative-free optimization methods, highlighting its scalability, robustness, and effectiveness in high-dimensional and complex optimization scenarios.

In the previous chapter, we presented a theoretical analysis of the sampling-and-classification (SAC) framework and identified two critical factors that influence its performance: the error-target dependence and the shrinking rate. The analysis revealed that these factors should be as small as possible to achieve better optimization efficiency. Inspired by these findings, we present a novel classification algorithm called the *randomized coordinate shrinking* algorithm, which aims to learn a discriminative model while keeping the error-target dependence and the shrinking rate small.

In this chapter, we introduce the randomized coordinate shrinking algorithm and its integration into the SAC framework, resulting in a new optimization algorithm called RACOS (RAndomized COordinate Shrinking) [14]. RACOS is designed to effectively optimize both continuous and discrete search spaces by leveraging the insights gained from the theoretical analysis. We conduct extensive experiments to compare RACOS with popular derivative-free optimization methods on various optimization benchmarks and machine learning tasks, including spectral clustering and classification with Ramp loss. The experimental results demonstrate the superiority of RACOS over the compared methods, highlighting its effectiveness and efficiency in solving complex optimization problems.

The rest of this chapter is organized as follows. Section 5.1 presents the randomized coordinate shrinking algorithm and its integration into the RACOS optimization

Y. Yu et al., *Derivative-Free Optimization*, Machine Learning: Foundations, Methodologies, and Applications, https://doi.org/10.1007/978-981-96-5929-6_5

algorithm. Section 5.2 describes the experimental setup and presents the empirical results on optimization testing functions, comparing RACOS with state-of-the-art derivative-free optimization algorithms. Section 5.3 evaluates RACOS on the spectral clustering task and compares its performance with classical clustering algorithms and evolutionary optimization methods. Section 5.4 investigates the effectiveness of RACOS on the classification task with Ramp loss and compares it with gradient-based and derivative-free optimization approaches. Finally, Sect. 5.5 summarizes the key findings and highlights the advantages of RACOS in high-dimensional and complex optimization problems.

5.1 The RACOS Optimization Algorithm

Theorem 4.3 in Chap. 4 has revealed that two critical factors, the error-target dependence and shrinking rate, should be as small as possible to achieve better optimization performance. However, these factors are not typically considered in traditional classification algorithms. Therefore, we need to design a new classification algorithm that explicitly takes these factors into account.

Algorithm 5.1 RACOS

Require:

f: Objective function to be minimized;
$\mathcal{C}$: A binary classification algorithm;
λ: Balancing parameter;
$T \in \mathbb{N}^+$: Number of iterations;
$m \in \mathbb{N}^+$: Sample size in each iteration;
$k \in \mathbb{N}^+ (\leq m)$: Number of positive samples;
`Sampling`: Sampling sub-procedure;
`Selection`: Decide the positive/negative solutions.

Ensure:

Collect $S_0 = \{\boldsymbol{x}_1, \ldots, \boldsymbol{x}_m\}$ by i.i.d. sampling from $\mathcal{U}_X$
$B_0 = \{(\boldsymbol{x}_1, y_1), \ldots, (\boldsymbol{x}_m, y_m)\}, \forall \boldsymbol{x}_i \in S_0 : y_i = f(\boldsymbol{x}_i)$
Let $(\tilde{\boldsymbol{x}}, \tilde{y}) = \arg\min_{(\boldsymbol{x},y)\in B^+} y$
for $t = 1$ to T **do**
  $(B_t^+, B_t^-) = \texttt{Selection}(B_{t-1}; k)$, $B_t = B_t^+$
  for $i = 1$ to m **do**
    $h_i = \mathcal{C}(B_t^+, B_t^-)$
    $\boldsymbol{x}_i = \begin{cases} \texttt{Sampling}(\mathcal{U}_{D_{h_i}}) & \text{w.p. } \lambda \\ \texttt{Sampling}(\mathcal{U}_X) & \text{w.p. } 1-\lambda \end{cases}$
    $y_i = f(\boldsymbol{x}_i)$ and let $B_t = B_t \cup \{(\boldsymbol{x}_i, y_i)\}$
  end for
  $(\tilde{\boldsymbol{x}}, \tilde{y}) = \arg\min_{(\boldsymbol{x},y)\in B_t \cup \{(\tilde{\boldsymbol{x}},\tilde{y})\}} y$
end for
return $(\tilde{\boldsymbol{x}}, \tilde{y})$

Inspired by the classical and simple *version space* learning algorithm [6], we present the *randomized coordinate shrinking* classification algorithm. Given a set of positive and negative solutions, the algorithm maintains an axis-parallel rectangle that covers all the positive solutions while excluding the negative ones. The learning process is highly randomized, and the rectangle is largely shrunk to meet the desired properties of small error-target dependence and shrinking rate.

The detailed steps of the proposed learning algorithm are depicted in Algorithm 5.2. The algorithm takes as input a set of solutions B_t with their corresponding objective values, which consists of positive and negative solutions according to a threshold α_t (cf. Algorithm 3.3 in Chap. 3). The goal is to discriminate a randomly selected positive solution (line 1) from the negative ones. In line 2, D_{h_t} denotes the positive region of the learned hypothesis h_t, and I is the index set of dimensions.

Algorithm 5.2 The *randomized coordinate shrinking* classification algorithm for $X = \{0, 1\}^n$ or $[0, 1]^n$

Require:

t: Current iteration number;
B_t^+, B_t^-: Positive and negative solution sets in iteration t;
X: Solution space ($\{0, 1\}^n$ or $[0, 1]^n$);
I: Index set of coordinates;
$M \in \mathbb{N}^+$: Maximum number of uncertain coordinates.

Ensure:

1: Randomly select $\boldsymbol{x}_+ = (x_+^{(1)}, \ldots, x_+^{(n)})$ from B_t^+
2: Let $D_{h_t} = X$, $I = \{1, \ldots, n\}$
3: **while** $\exists \boldsymbol{x} \in B_t^-$ s.t. $h_t(\boldsymbol{x}) = +1$ **do**
4:   **if** $X = \{0, 1\}^n$ **then**
5:     k = randomly selected index from the index set I
6:     $D_{h_t} = D_{h_t} - \{\boldsymbol{x} \in X \mid x^{(k)} \neq x_+^{(k)}\}$, $I = I - \{k\}$
7:   **end if**
8:   **if** $X = [0, 1]^n$ **then**
9:     k = randomly selected index from the index set I
10:     $\boldsymbol{x}^-$ = randomly selected solution from B_t^-
11:     **if** $x_+^{(k)} \geq x_-^{(k)}$ **then**
12:       r = uniformly sampled value in $(x_-^{(k)}, x_+^{(k)})$
13:       $D_{h_t} = D_{h_t} - \{\boldsymbol{x} \in X \mid x^{(k)} < r\}$
14:     **else**
15:       r = uniformly sampled value in $(x_+^{(k)}, x_-^{(k)})$
16:       $D_{h_t} = D_{h_t} - \{\boldsymbol{x} \in X \mid x^{(k)} > r\}$
17:     **end if**
18:   **end if**
19: **end while**
20: **while** $|I| > M$ **do**
21:   k = randomly selected index from the index set I
22:   $D_{h_t} = D_{h_t} - \{\boldsymbol{x} \in X \mid x^{(k)} \neq x_+^{(k)}\}$, $I = I - \{k\}$
23: **end while**
24: **return** h_t

The algorithm consists of two main steps: learning with randomness until all negative solutions have been excluded (lines 3–19) and shrinking (lines 20–23). In the learning step, we consider both discrete ($X = \{0, 1\}^n$) and continuous ($X = [0, 1]^n$) domains. For the discrete domain (lines 4–7), the algorithm randomly selects a dimension and collapses that dimension to the value of the positive solution (lines 5–6). For the continuous domain (lines 8–18), the algorithm sets the upper or lower bound on a randomly chosen dimension to exclude negative solutions (lines 11–17). This process can be easily extended to larger vocabulary sets or general box constraints. Finally, lines 20–23 further shrink the classifier to leave only M dimensions uncollapsed, applicable to both discrete and continuous domains. This learning algorithm with high-level randomness achieves a positive region with a small error-target dependence and largely reduces the positive region for a small shrinking rate.

By incorporating this classification algorithm into Algorithm 5.1 (implementing the $\mathcal{C}$ in line 7), we obtain the RACOS optimization algorithm. Note that the `Sampling` procedure of Algorithm 3.3 (line 8) simply draws a solution uniformly from the rectangular positive area defined by the learned hypothesis. The `Selection` procedure simply select the top k solutions to be positive.

5.2 Empirical Study on Testing Functions

We first empirically evaluate RACOS on minimizing two benchmark testing functions: the convex Sphere function and the highly non-convex Ackley function, defined as follows:

$$\text{Sphere:} \quad f(\boldsymbol{x}) = \sum_{i=1}^{n} (x_i - 0.2)^2, \tag{5.1}$$

$$\text{Ackley:} \quad f(\boldsymbol{x}) = -20e^{\left(-\frac{1}{5}\sqrt{\frac{1}{n}\sum_{i=1}^{n}(x_i-0.2)^2}\right)} - e^{\left(\frac{1}{n}\sum_{i=1}^{n}\cos 2\pi x_i\right)} + e + 20. \tag{5.2}$$

The goal is to minimize these functions within the solution space $X = [0, 1]^n$, where the global minimum values are 0. Note that although the two functions look quite different, they both satisfy local Lipschitz continuity. We expect RACOS performs good on the two functions, particularly on the sophisticated Ackley function.

We compare RACOS with the following state-of-the-art derivative-free optimization algorithms:

- Simultaneous Optimistic Optimization (SOO) [8, 9].
- Random Embedding Bayesian Optimization (REMBO) [13].
- Covariance Matrix Adaptation Evolution Strategy (CMA-ES) [5].

The implementations of these algorithms are provided by their respective authors.

To study the scalability of the algorithms with respect to the dimensionality of the solution space, we set $n \in \{10, 100, 500, 1000\}$ and fix the maximum number of function evaluations to $30n$ for all algorithms. To investigate the convergence rate

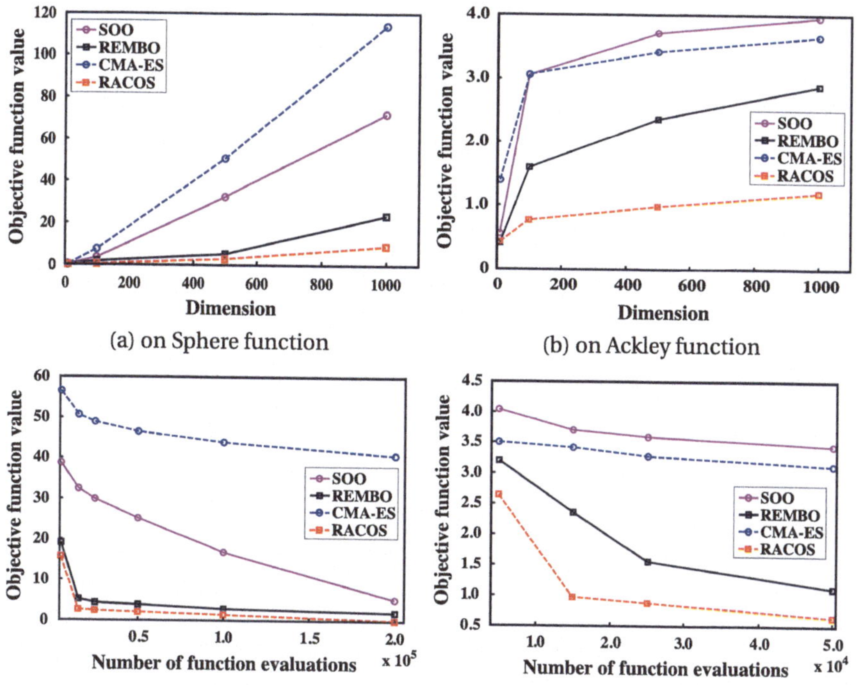

(a) on Sphere function (b) on Ackley function (c) on Sphere function (d) on Ackley function

Fig. 5.1 Comparing the scalability with $30n$ evaluations in **a** and **b**, and the convergence rate with $n = 500$ in **c** and **d** [14]

with respect to the number of function evaluations, we choose $n = 500$ and vary the total number of function evaluations from 5×10^3 to 2×10^5 for the Sphere function and from 5×10^3 to 5×10^4 for the Ackley function.

Each algorithm is run independently for 30 times, and the mean of the achieved objective values is reported in Fig. 5.1.

Figure 5.1a, b show that RACOS achieves the lowest growth rate as the dimension increases, indicating its superior scalability compared to the other algorithms. Figure 5.1c, d demonstrate that RACOS reduces the objective function value at the highest rate, implying its consistently faster convergence than the other methods.

5.3 Empirical Study on Clustering Task

Next, we evaluate RACOS on a kind of machine learning task, i.e., clustering task. Specifically, we consider the RatioCut problem. Given a dataset $\mathcal{V} = \{\boldsymbol{v}_1, \ldots, \boldsymbol{v}_n\}$, the goal is to cluster the data points into two groups, $\{A_1, A_2\}$, by minimizing the inter-cluster similarity. The solution space is naturally represented by the discrete

Table 5.1 Comparing the achieved objective values of the algorithms (mean ± standard deviation) [14]. In each column, the entry with the best (smallest) mean value is bolded. An entry marked with a bullet (•) indicates that it is significantly worse than the best algorithm according to a t-test with a confidence level of 5%. The last column counts the win/tie/loss of each algorithm compared to RACOS

Algo.	*Sonar*	*Heart*	*Ionosphere*	*Breast Cancer*	*German*	w/t/l to RACOS
USC	3.91±0.00•	79.67±0.00•	54.21±0.00•	200.62±0.00•	239.00±0.00•	0/0/5
GA	3.14±0.74	**57.31**±0.46	55.71±3.74•	189.52±1.26	205.61±1.80•	0/3/2
RLS	4.07±0.82•	58.81±0.45•	58.74±2.81•	192.63±1.62•	207.36±2.11•	0/0/5
UMDA	7.40±2.26•	58.76±1.02•	61.77±4.54•	193.58±3.56•	212.83±1.08•	0/0/5
CE	8.00±1.35•	58.75±1.39•	63.71±3.41•	188.76±3.77	209.57±1.96•	0/1/4
RACOS	**2.88**±0.63	57.45±0.89	**50.01**±2.80	**187.55**±3.01	**192.11**±2.51	-/-/-

domain $X = \{0, 1\}^n$ for the bipartition. The optimization objective is formulated as

$$f(A_1, A_2) = \sum_{i=1}^{2} \frac{1}{|A_i|} \sum_{p \in A_i, q \notin A_i} W_{p,q} \text{ over } X, \tag{5.3}$$

where $W_{p,q} = \exp(-\|\boldsymbol{v}_p - \boldsymbol{v}_q\|_2^2/\sigma^2)$ is the similarity between $\boldsymbol{v}_p$ and $\boldsymbol{v}_q$. The RatioCut problem is known to be NP-hard.

We compare RACOS with the following algorithms:

- Unnormalized Spectral Clustering (USC) [12]: a classical approximate algorithm for the RatioCut problem.
- Genetic Algorithm (GA) [4]: using bit-wise mutation with probability $1/n$ and one-bit crossover with probability 0.5.
- Randomized Local Search (RLS) [10]
- Univariate Marginal Distribution Algorithm (UMDA) [7]
- Cross-Entropy (CE) method [3].

The parameters for GA, RLS, UMDA, and CE are set according to the recommendations in their respective references.

We use five binary UCI datasets [1]: *Sonar*, *Heart*, *Ionosphere*, *Breast Cancer*, and *German*, with 208, 270, 351, 683, and 1000 instances, respectively. All features are normalized into the range $[-1, 1]$.

The total number of calls to the objective function for GA, RLS, UMDA, CE, and RACOS is set to $30n$. Each algorithm is run independently for 30 times on each dataset. Table 5.1 reports the achieved objective values.

Table 5.1 shows that, according to a t-test with a confidence level of 5%, RACOS is never worse than the other algorithms. It consistently outperforms USC, RLS, and UMDA, and achieves significant wins over GA and CE. These results demonstrate that RACOS not only exhibits superior performance but also maintains stability across different datasets.

5.4 Empirical Study on Classification with Ramp Loss

We further evaluate RACOS on another kind of machine learning task, i.e., classification task with the Ramp loss function [2]. The Ramp loss is defined as

$$R_s(z) = H_1(z) - H_s(z) \text{ with } s < 1, \tag{5.4}$$

where $H_s(z) = \max\{0, s - z\}$ is the Hinge loss with s being the hinge point. The objective is to find a vector $\boldsymbol{w}$ and a scalar b that minimize:

$$f(\boldsymbol{w}, b) = \frac{1}{2}\|\boldsymbol{w}\|_2^2 + C\sum_{\ell=1}^{L} R_s\big(y_\ell(\boldsymbol{w}^\top \boldsymbol{v}_\ell + b)\big), \tag{5.5}$$

where $\boldsymbol{v}_\ell$ is the ℓth training instance and $y_\ell \in \{-1, +1\}$ is its corresponding label. This objective function is similar to that of Support Vector Machines (SVM) [11], but SVM uses the Hinge loss instead. Due to the convexity of the Hinge loss, the number of support vectors in SVM increases linearly with the number of training instances, which can be undesirable in terms of scalability. This issue can be alleviated by using the Ramp loss [2].

We compare RACOS with the following algorithms:

- Simultaneous Optimistic Optimization (SOO) [8, 9]
- Random Embedding Bayesian Optimization (REMBO) [13]
- Covariance Matrix Adaptation Evolution Strategy (CMA-ES) [5]
- Concave-Convex Procedure (CCCP) [16]: a gradient-based non-convex optimization approach for objective functions that can be decomposed into a convex sub-function plus a concave sub-function.

We use two binary-class UCI datasets, *Adult* and *USPS+N* (0 vs. rest), which were also used in [2]. The feature dimensions of these datasets are 123 and 256, respectively. All features are normalized into the range [0, 1] or [−1, 1]. Since our focus is on optimization performance, we compare the results on the training set.

To study the effectiveness of RACOS under different hyper-parameter settings, we test $s \in \{-1, 0\}$ and $C \in \{0.1, 0.5, 1, 2, 5, 10\}$, as there are two hyper-parameters (C and s) in the optimization formulation.

The total number of calls to the objective function is set to $40n$ for all algorithms except CCCP, while CCCP runs until convergence. Each algorithm is run independently for 30 times. The achieved objective values are reported in Fig. 5.2.

As shown in Fig. 5.2, RACOS consistently achieves the best performance compared to SOO, REMBO, and CMA-ES in all settings. It is worth noting that the optimization difficulty increases with C, as a smaller C corresponds to an objective function closer to being convex. On the *USPS+N* dataset, we observe that CCCP performs the best when the objective function is very close to being convex (i.e., when C is very small), which can be attributed to its gradient-based nature. However, CCCP does

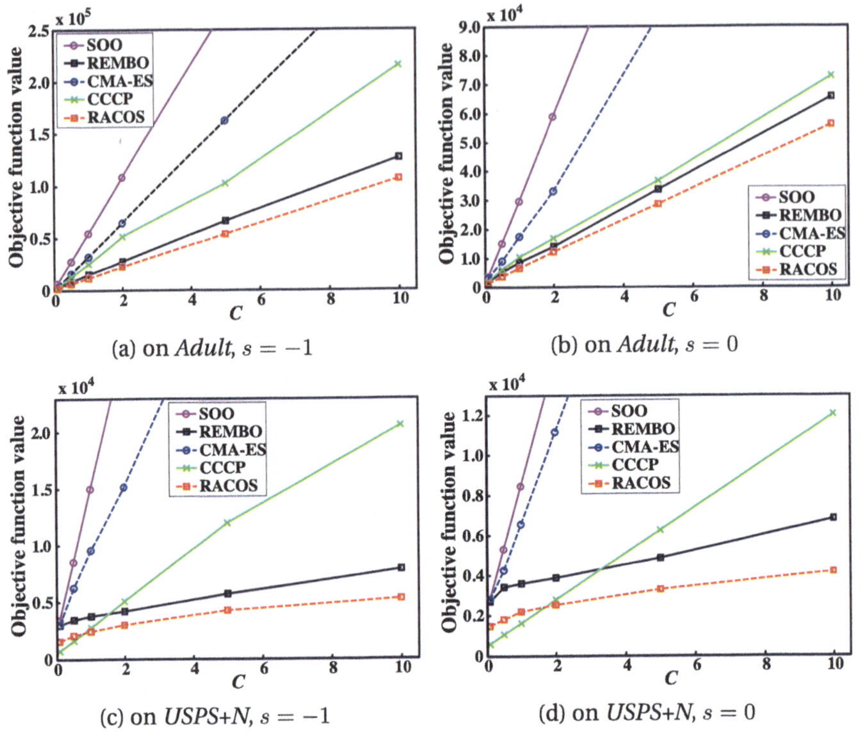

Fig. 5.2 Comparing the achieved objective function values against the parameter C of the classification with Ramp loss [14]

not perform well in highly non-convex scenarios. Furthermore, the advantage of RACOS becomes more pronounced as C increases in all situations, indicating its suitability for complex optimization tasks.

5.5 Summary

In this chapter, we introduced the *randomized coordinate shrinking* algorithm for learning the classification model, inspired by the critical factors identified in Chap. 4. By integrating this algorithm into the SAC framework, we presented the RACOS [15] optimization algorithm, which is applicable to both continuous and discrete search spaces. Experimental results on optimization benchmarks and machine learning tasks, including spectral clustering and classification with Ramp loss, demonstrated the superiority of RACOS compared to other state-of-the-art optimization methods.

Moreover, we observed that RACOS exhibits better scalability in high-dimensional optimization problems and maintains a clear advantage as the difficulty of the optimization problem increases. These findings highlight the effectiveness and robustness of RACOS in tackling complex and challenging optimization tasks.

The success of RACOS can be attributed to its ability to learn a discriminative model while keeping the error-target dependence and the shrinking rate small, as suggested by the theoretical analysis in Chap. 4. The randomized coordinate shrinking algorithm effectively balances the trade-off between exploration and exploitation, enabling RACOS to efficiently navigate the search space and converge to high-quality solutions.

In the next chapter, we will further extend the RACOS algorithm to handle optimization problems with mixed continuous and discrete variables, as well as those with black-box constraints. We will also explore the potential of integrating RACOS with other optimization techniques to develop more powerful and versatile optimization algorithms.

References

1. Blake CL, Keogh E, Merz CJ (1998) UCI Repository of machine learning databases. http://www.ics.uci.edu/~mlearn/MLRepository.html
2. R. Collobert, F. Sinz, J. Weston, and L. Bottou. Trading convexity for scalability. In *Proceedings of the 23rd International Conference on Machine learning*, pages 201–208, Pittsburgh, Pennsylvania, 2006
3. de Boer P, Kroese DP, Mannor S, Rubinstein RY (2005) A tutorial on the cross-entropy method. Ann Oper Res 134(1):19–67
4. Goldberg D (1989) Genetic Algorithms in Search, Optimization, and Machine Learning. Addison-Wesley, Reading, MA
5. Hansen N, Müller SD, Koumoutsakos P (2003) Reducing the time complexity of the derandomized evolution strategy with covariance matrix adaptation (CMA-ES). Evol Comput 11(1):1–18
6. T. Mitchell. *Machine Learning*. McGraw Hill, 1997
7. Mühlenbein H (1997) The equation for response to selection and its use for prediction. Evol Comput 5(3):303–346
8. R. Munos. Optimistic optimization of a deterministic function without the knowledge of its smoothness. In *Advances in Neural Information Processing Systems 24*, pages 783–791, Granada, Spain, 2011
9. Munos R (2014) From bandits to Monte-Carlo tree search: The optimistic principle applied to optimization and planning. Foundations and Trends in Machine Learning 7(1):1–130
10. Neumann F, Wegener I (2007) Randomized local search, evolutionary algorithms, and the minimum spanning tree problem. Theoret Comput Sci 378(1):32–40
11. V. Vapnik. *The Nature of Statistical Learning Theory*. Springer Science & Business Media, 2000
12. Von Luxburg U (2007) A tutorial on spectral clustering. Stat Comput 17(4):395–416
13. Z. Wang, M. Zoghi, F. Hutter, D. Matheson, and N. De Freitas. Bayesian optimization in high dimensions via random embeddings. In *Proceedings of the 23rd International Joint Conference on Artificial Intelligence*, pages 1778–1784, Beijing, China, 2013
14. Y. Yu, H. Qian, and Y. Hu. Derivative-free optimization via classification. In *Proceedings of the 30th AAAI Conference on Artificial Intelligence*, pages 2286–2292, Phoenix, Arizona, 2016a
15. Y. Yu, H. Qian, and Y.-Q. Hu. Derivative-free optimization via classification. In *Proceedings of the 30th AAAI Conference on Artificial Intelligence*, pages 2286–2292, Phoenix, AZ, 2016b

16. A. L. Yuille and A. Rangarajan. The concave-convex procedure (CCCP). In *Advances in Neural Information Processing Systems 14*, pages 1033–1040, Vancouver, Canada, 2001

Part III
Practical Extensions

Chapter 6
Optimization in Sequential Mode

Abstract This chapter introduces SRacos, a sequential-mode classification-based derivative-free optimization method designed to address the limitations of batch-mode optimization in scenarios where samples and their evaluations must be obtained sequentially. Unlike batch-mode methods, which require a set of samples to update the model, SRacos updates the sampling model immediately after evaluating each sample by reusing historical data from previous iterations. This approach improves sample efficiency, requiring fewer samples to achieve the same optimization goal compared to batch-mode methods. The chapter provides a theoretical analysis of SRacos, demonstrating its potential for better query complexity under certain conditions. Empirical studies compare SRacos with state-of-the-art optimization algorithms, including CMA-ES, DE, CE, and IMGPO, on synthetic functions and reinforcement learning tasks. Results show that SRacos consistently outperforms batch-mode methods in convergence rate and scalability, particularly in high-dimensional and complex optimization problems. The chapter concludes by highlighting the advantages of sequential-mode optimization in accelerating the optimization process.

In the previous chapters, we introduced the sampling-and-learning (SAL) and sampling-and-classification (SAC) frameworks for derivative-free optimization and proposed the RACOS algorithm, which operates in a batch mode. However, in certain real-world scenarios, obtaining a batch of samples and their evaluations simultaneously may not be feasible or efficient. For example, in reinforcement learning tasks solved by direct policy search, each policy sample relies on the previous samples, and the optimization process can only obtain the samples and their evaluations sequentially. Another example is solving AutoML tasks using derivative-free optimization, where the evaluation process is often extremely expensive, making it difficult to provide a batch of evaluations for the optimization process. In such sequential situations, the batch-mode derivative-free optimization methods may be inefficient.

To address this challenge, we present a sequential-mode classification-based derivative-free optimization method called SRACOS [4]. SRACOS aims to improve the optimization efficiency by updating the sampling model immediately after evaluating each sample. However, updating the model typically requires a batch of samples according to the original batch-mode optimization framework, and a single

Y. Yu et al., *Derivative-Free Optimization*, Machine Learning: Foundations, Methodologies, and Applications, https://doi.org/10.1007/978-981-96-5929-6_6

sample cannot complete the update process. To overcome this issue, SRACOS considers reusing the samples from previous iterations. Through theoretical analysis, we prove that SRACOS can achieve better sample efficiency compared to the batch-mode method, meaning that it requires fewer samples to achieve the same optimization goal.

6.1 Sequential Classification Model Based Algorithm

It is important to note that RACOS operates in a batch mode: the hypothesis h_i depends on B_t^+ and B_t^- in line 7 of Algorithm 5.1, and within the loop of iteration (lines 6–10), those two sets remain unchanged. This means that the sampling regions are generated from the same distribution, even if this distribution is not optimal. Batch-mode sampling may produce redundancy and may not be suitable for sequential-mode problems like direct policy search. To address this limitation, we present the sequential RACOS (SRACOS) algorithm.

Building upon RACOS, a straightforward idea for converting RACOS to a sequential-mode algorithm is to update the sets B^+ and B^- immediately after obtaining a sample and its evaluation value. However, in this case, there is only one new sample, which cannot replace the entire B^+ and B^- sets as RACOS does. To tackle this problem, the sequential-mode algorithm reuses the historical samples from previous iterations, resulting in the SRACOS algorithm. The pseudo-code of SRACOS is shown in Algorithm 6.1.

In Algorithm 6.1, `Sampling` $(\mathcal{S})$ denotes a sample sub-procedure which obtains a sample from a given distribution $\mathcal{S}$. `Selection` $(B; k)$ denotes a sub-procedure

Algorithm 6.1 Sequential RACOS (SRACOS)

Require: (extra input than RACOS)
$N \in \mathbb{N}^+$: Budget;
$r = m + k$;
`Replace`: Replacing sub-procedure.
Ensure:
Collect $S = \{\boldsymbol{x}_1, \ldots, \boldsymbol{x}_r\}$ by i.i.d. sampling from $\mathcal{U}_X$
$B = \{(\boldsymbol{x}_1, y_1), \ldots, (\boldsymbol{x}_r, y_r)\}, \forall \boldsymbol{x}_i \in S : y_i = f(\boldsymbol{x}_i)$
$(B^+, B^-) =$ `Selection`$(B; k)$
Let $(\tilde{\boldsymbol{x}}, \tilde{y}) = \text{argmin}_{(\boldsymbol{x},y)\in B^+} y$
for $t = r + 1$ to N **do**
  $h = \mathcal{C}(B^+, B^-)$
  $\boldsymbol{x} = \begin{cases} \texttt{Sampling}(\mathcal{U}_{D_h}) & \text{w.p. } \lambda \\ \texttt{Sampling}(\mathcal{U}_X) & \text{w.p. } 1 - \lambda \end{cases}$
  $y = f(\boldsymbol{x})$
  $[(\boldsymbol{x}', y'), B^+] =$ `Replace`$((\boldsymbol{x}, y), B^+,$ 'strategy_P')
  $[\emptyset, B^-] =$ `Replace`$((\boldsymbol{x}', y'), B^-,$ 'strategy_N')
  $(\tilde{\boldsymbol{x}}, \tilde{y}) = \arg\min_{(\boldsymbol{x},y)\in B^+ \cup \{(\tilde{\boldsymbol{x}}, \tilde{y})\}} y$
end for
return $(\tilde{\boldsymbol{x}}, \tilde{y})$

which splits the sample-evaluation pair set B into two subsets B^+ and B^-, where B^+ contains the samples which has top-k best evaluation values. `Replace` $(a, A, \text{'}s\text{'})$ is a replacing sub-procedure that uses a to replace a sample-evaluation value pair from a set A with a strategy "s" and update the pair set A. The return of this sub-procedure is a replaced pair. Three replacing strategies considered are as follows: replacing the pair with worst evaluation value in A (WR-); replacing a pair in A randomly (RR-) and replacing the pair which has largest margin from the best-so-far solution (LM-).

In Algorithm 6.1, after initialization, SRACOS will get two pair sets B^+ and B^-, denoting the positive sample-value pair set and the negative sample-value pair set. The method of generating a new sample (lines 7 to 8) is the same as RACOS. After getting a new pair $(\boldsymbol{x}, y)$, SRACOS updates B^+ and B^- immediately. When updating B^+ (line 9), SRACOS uses the "strategy_P". Because B^+ must contain the best-so-far samples, "strategy_P" can only be "WR-", i.e., a sample with the worst evaluation value is removed from $B^+ \cup \{(\boldsymbol{x}, y)\}$ and the rest of set is the new B^+. The removed pair $(\boldsymbol{x}', y')$ is used to update B^- using "strategy_N" in line 10. "strategy_N" can be any one of three strategies mentioned above. In experiments section, we will prove that selection of "strategy_N" has no influence on convergence rate of SRACOS empirically. In the end, SRACOS returns the best sample-value pair $(\tilde{x}, \tilde{y})$.

6.2 Theoretical Analysis

In this section, we analyze the (ϵ,δ)-query complexity of the sequential classification-based optimization algorithm SRACOS. We adopt the same notations and definitions introduced in Chap. 4, including the error-target θ-dependence (Definition 4.3) and γ-shrinking rate (Definition 4.4).

For simplicity, let $t = r + 1, \ldots, N$. We derive an upper bound on the (ϵ,δ)-query complexity of SRACOS under the conditions of error-target θ-dependence and γ-shrinking rate.

Theorem 6.1 *Given $f \in \mathcal{F}$, $0 < \delta < 1$, and $\epsilon > 0$, if* SRACOS *has error-target θ-dependence and γ-shrinking rate, then its (ϵ,δ)-query complexity is upper bounded by*

$$O\left(\max\left\{\frac{1}{\mu(D_\epsilon)}\left((1-\lambda) + \frac{\lambda}{\gamma(N-r)}\sum_{t=r+1}^{N}\Phi_t\right)^{-1}\ln\frac{1}{\delta},\ N\right\}\right),$$

where $\Phi_t = \left(1 - \epsilon_{\mathcal{D}_t} - \sqrt{2D_{KL}(\mathcal{D}_t \| \mathcal{U}_X)} - \theta\right) \cdot \mu(D_{\alpha_t})^{-1}$ and $|X|$ is the volume of X.

Proof By Lemma 4.1 (Chap. 4), we have $\Pr_{h_t} \geq \mu(D_\epsilon \cap D_{h_t})/\mu(D_{h_t})$ for all t. Combining Lemma 4.3 (Chap. 4) with Lemma 4.4 (Chap. 4), we can conclude that

$$\mu(D_\epsilon \cap D_{h_t}) \geq \mu(D_\epsilon) \cdot (1 - \epsilon_{\mathcal{D}_t} - \sqrt{2D_{KL}(\mathcal{D}_t \| \mathcal{U}_X)} - \theta),$$

where $\mathcal{D}_t$ is the true sampling distribution on which h_t is learned. Unlike the batch-mode, where the distribution is a combination of uniform sampling and sampling from the model, i.e., $\mathcal{D}_t = \lambda \mathcal{U}_{D_{h_t}} + (1-\lambda)\mathcal{U}_X$, the distribution in the sequential model is a combination of uniform sampling and sampling from a model set H determined by the strategy of keeping previous samples, i.e., $\mathcal{D}_t = \lambda \frac{1}{|H_t|} \sum_{h \in H_t} \mathcal{U}_{D_h} + (1-\lambda)\mathcal{U}_X$. For generality, we keep using the notation $\mathcal{D}_t$ without specifying it.

Meanwhile, the γ-shrinking rate condition directly admits $\mu(D_{h_t}) \leq \gamma \mu(D_{\alpha_t})$ for all t. Let $\Phi_t = \left(1 - \epsilon_{\mathcal{D}_t} - \sqrt{2D_{KL}(\mathcal{D}_t \| \mathcal{U}_X)} - \theta\right) \cdot \mu(D_{\alpha_t})^{-1}$. Therefore, $\Pr_{h_t} \geq \gamma^{-1}\mu(D_\epsilon)\Phi_t$. On the other hand, by the procedure of SRacos, we have $\sum_{t=1}^{T} m_{\Pr_{h_t}} = \sum_{t=r+1}^{N} m_{\Pr_{h_t}} \in O(N)$. Finally, by the definition of $\overline{\Pr}_h$ and Theorem 4.1 (Chap. 4), we prove the theorem. □

To explicitly compare the query complexity of SRacos with the batch-mode Racos (whose query complexity upper bound is shown in Theorem 4.3 of Chap. 4), we have the following theorem:

Theorem 6.2 *Ignoring the constant factor and fixing θ and γ,* SRacos *can have a better (or worse) query complexity upper bound than* Racos *if for any iteration t*

$$\epsilon_{\mathcal{D}_t^S} < (or >) \frac{1}{1-\lambda}\epsilon_{\mathcal{D}_t^B} - \sqrt{2D_{KL}(\mathcal{D}_t^S \| \mathcal{U}_X)},$$

where $\mathcal{D}_t^S$ and $\mathcal{D}_t^B$ denote the distributions under which the classifier is trained at iteration t of SRacos *and* Racos, *respectively, and $\epsilon_{\mathcal{D}_t^S}$ and $\epsilon_{\mathcal{D}_t^B}$ denote their corresponding generalization errors.*

Proof In Theorem 6.1, ignoring the constant factor and letting $\epsilon > 0$ be small enough such that we only need to focus on the term

$$\frac{1}{\mu(D_\epsilon)}\left((1-\lambda) + \frac{\lambda}{\gamma(N-r)}\sum_{t=r+1}^{N}\Phi_t\right)^{-1} \ln\frac{1}{\delta},$$

where $\Phi_t = \left(1 - \epsilon_{\mathcal{D}_t^S} - \sqrt{2D_{KL}(\mathcal{D}_t^S \| \mathcal{U}_X)} - \theta\right) \cdot \mu(D_{\alpha_t})^{-1}$.

Based on Theorems 6.1 and 4.3, to compare SRacos with Racos, it suffices to compare the term $1 - \epsilon_{\mathcal{D}_t^S} - \sqrt{2D_{KL}(\mathcal{D}_t^S \| \mathcal{U}_X)} - \theta$ with $1 - (1-\lambda)^{-1}\epsilon_{\mathcal{D}_t^B} - \theta$, ignoring the corresponding constant factors. It can be verified directly that, for any iteration t, if $\epsilon_{\mathcal{D}_t^S} < (1-\lambda)^{-1}\epsilon_{\mathcal{D}_t^B} - \sqrt{2D_{KL}(\mathcal{D}_t^S \| \mathcal{U}_X)}$, then SRacos has a better query complexity upper bound than Racos; if $\epsilon_{\mathcal{D}_t^S} > (1-\lambda)^{-1}\epsilon_{\mathcal{D}_t^B} - \sqrt{2D_{KL}(\mathcal{D}_t^S \| \mathcal{U}_X)}$, then SRacos is worse. □

Theorem 6.2 implies that when γ is close to 1, i.e., more exploitation than exploration, the sequential mode can be often better than the batch mode.

6.3 Empirical Study

In this section, we conduct experiments to investigate the effectiveness of SRACOS. We compare SRACOS with state-of-the-art methods, including CMA-ES [2], differential evolution algorithm (DE) [8], cross-entropy method (CE), and a Bayesian optimization method with exponential convergence (IMGPO) [5]. In our experiments, all these algorithms use their default hyper-parameter settings.

We select two types of tasks: optimization on two synthetic testing functions and direct policy search for reinforcement learning, including the helicopter hovering control task [6].

6.3.1 *Optimization on Synthetic Functions*

We select two benchmark testing functions: the convex Sphere function,

$$\text{Sphere:} \quad f(\boldsymbol{x}) = \sum_{i=1}^{n}(x_i - 0.2)^2,$$

and the highly non-convex Ackley function,

$$\text{Ackley:} \quad f(\boldsymbol{x}) = -20e^{(-\frac{1}{5}\sqrt{\sum_{i=1}^{n}(x_i-0.2)^2})} - e^{\frac{1}{n}\sum_{i=1}^{n}\cos 2\pi(x_i-0.2)} + e + 20.$$

The goal is to minimize both functions within the search space $X = [-1, 1]^n$. The optimal values of both functions are 0 at the optimal solution $\boldsymbol{x}^* = \{0.2\}^n$. The dimensionality of the functions is set to $n = 100$ and 1000. For the optimization process, we set the number of evaluated samples to $20n$. We run each experiment independently for 15 times and report the average performance.

First, we investigate the effectiveness of the selection of "strategy_N" for SRACOS. The compared methods in this experiment are SRACOS with "WR-", "RR-", "LM-" strategies and the batch-mode algorithm RACOS. The results are shown in Fig. 6.1. The convergence curves of SRACOS with three replacing strategies almost overlap, indicating that we can select any of the three replacing strategies in practice. In the rest of the experiments, we select "WR-" as the replacing strategy.

Next, we focus on the convergence rate of all compared methods. The results are shown in Fig. 6.2. Comparing SRACOS with RACOS, it is clear that SRACOS consistently outperforms RACOS in all experiment settings. In low dimensionality and convex function ($n = 100$, Sphere), the Bayesian optimization (IMGPO) shows the highest convergence rate, but it struggles with high-dimensional problems ($n = 1000$) or problems with many local optima (Ackley). In contrast, SRACOS demonstrates the best convergence rate in those settings.

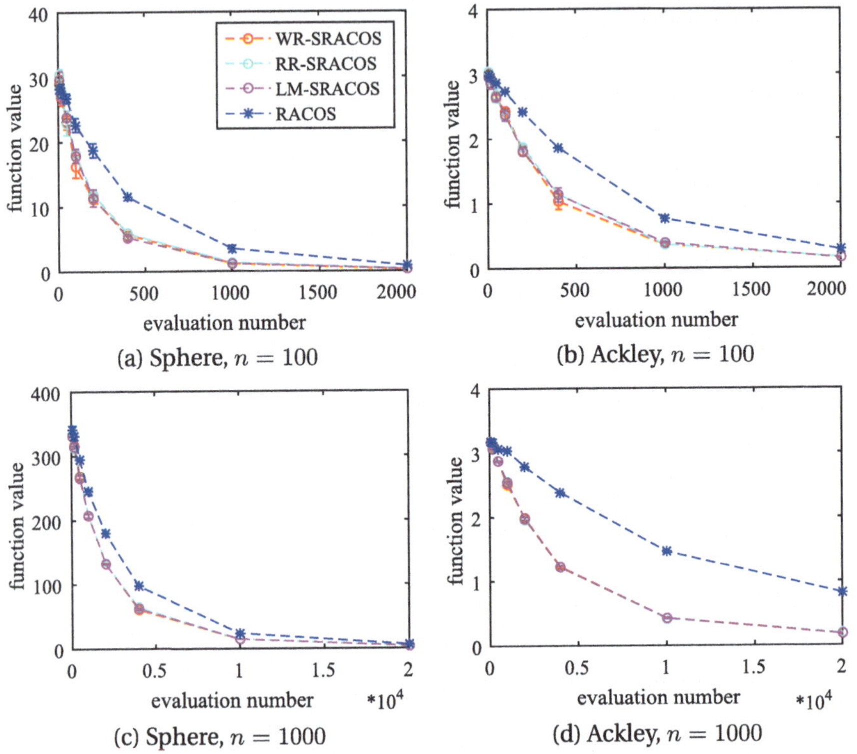

Fig. 6.1 The effectiveness investigation of "strategy_N" of SRACOS on Sphere and Ackley functions [4]

The results of the scalability of compared methods are shown in Fig. 6.3. The dimensionality is set as $n = 10, 20, 50, 100, 200, 500, 1000, 2000$. The total evaluation number for each dimension setting is $20n$. SRACOS shows the lowest performance growth rate as the dimensionality increases, indicating that SRACOS has the best scalability among all compared methods. We also consider the real wall-clock time that the compared methods cost and show the results in Fig. 6.4. SRACOS takes more computation time than RACOS. However, SRACOS achieves better final performance than RACOS. Even with the same wall-clock time, SRACOS outperforms RACOS. All results verify that the sequential mode can effectively accelerate the optimization process.

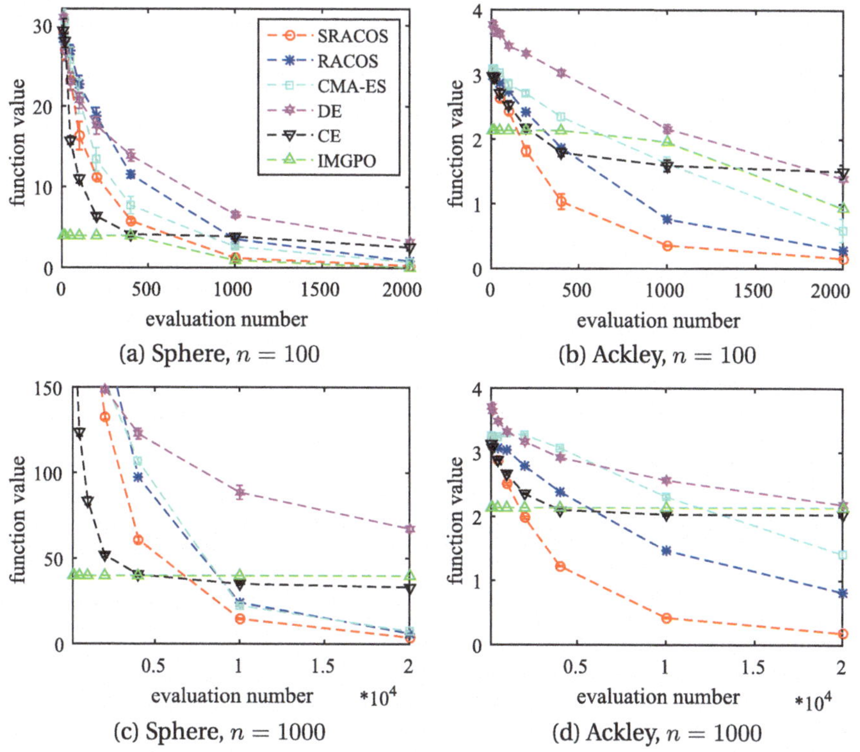

Fig. 6.2 The convergence rate of compared methods on Sphere and Ackley functions with dimensionality $n = 100$ and $n = 1000$ [4]

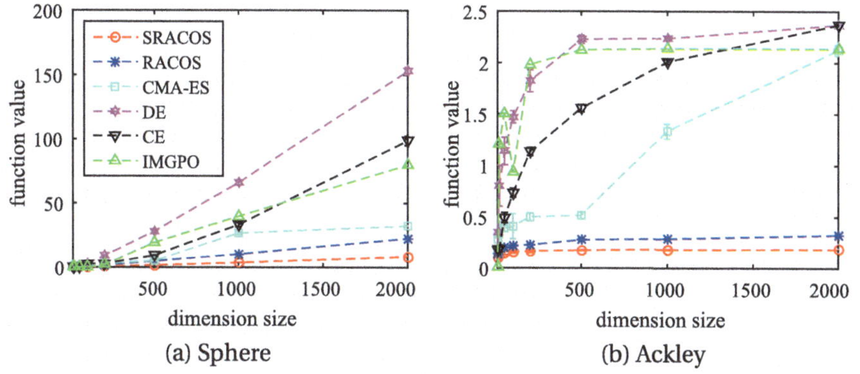

Fig. 6.3 The scalability of compared methods on Sphere and Ackley functions [4]

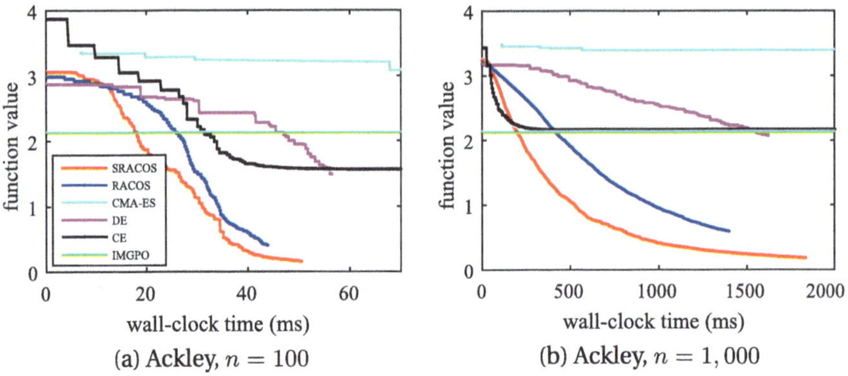

Fig. 6.4 The convergence speed against the wall-clock time on Ackley functions with $20n$ evaluations [4]

6.3.2 *Direct Policy Search on Reinforcement Learning Tasks*

In this section, we use derivative-free optimization to solve direct policy search for reinforcement learning tasks. Reinforcement learning can be described by a Markov decision process (MDP) which consists of a tuple $\langle S, A, P_{sa}, R\rangle$. S is the state space, A is the action space, P_{sa} is the transition function which indicates the next state when taking action a at state s, and $R : S \to \mathbb{R}$ is the reward function which gives the reward when taking an action. The dynamics of MDP are as follows: the environment initializes a state s_0, then the policy chooses an action a_0 based on s_0. The environment transitions to the next state s_1 according to the transition function $P_{s_0a_0}$ and provides a reward r_0. Then, the policy takes the action a_1 based on s_1, and so on. The sequence of $s_0, s_1, \ldots, s_N$ denotes a trajectory. Let $\pi : S \to A$ denote the policy. The goal of reinforcement learning is to maximize the long-term accumulated reward: $\sum_{i=1}^{N} r_i$. In the experiments, we select a feed-forward fully connected neural network as the policy. Thus, the objective is to optimize the weights of the policy network to maximize the accumulated reward.

Helicopter Hovering Control Task. Helicopter flight is regarded as a challenging control task that has been successfully solved by reinforcement learning [6]. In this task, the helicopter should be controlled to stay in a hovering state within a limited region. If the helicopter moves out of the region, it is considered crashed, and the policy will receive a very low reward. Previously, neural networks have been proven to be a good policy model [7]. In this experiment, we select a fully connected neural network without any hidden layer as the policy. The environment flies the helicopter for a total of 2000 steps. The sum of rewards is the evaluation of the policy. Let $\boldsymbol{w}$ denote the weights of the policy. We set $\boldsymbol{w} \in [-10, 10]^n$ as the policy search space. The helicopter environment has a 13-dimensional state space and a 4-dimensional action space. Thus, the policy has a total of 52 weights. The process of direct policy search can be presented as follows: the derivative-free optimization methods generate a set of weights as the policy, the environment flies the helicopter under the control of this policy, and the sum of rewards is the evaluation value for this policy. The derivative-free optimization method will generate a new set of weights based on the reward feedback. We compare SRACOS with RACOS, CMA-ES, DE, CE, and IMGPO. We run each experiment for 15 times for each algorithm. The evaluation number is 10^5 for all derivative-free optimization methods.

We show the convergence curves of the reward and hovering steps in Fig. 6.5. The average performance is shown in Table 6.1. The results demonstrate that SRACOS can find the best policy faster than other compared methods, indicating that SRACOS has the fastest optimization convergence rate in the helicopter hovering control task.

Gym Tasks. We select 8 OpenAI Gym control tasks, including: “Acrobot”, “MountainCar”, “HalfCheetah”, “Humanoid”, “Swimmer”, “Ant”, “Hopper”, and “LunarLander”. A fully connected neural network is again used as the control policy. Due to the different state and action spaces of the tasks, we set different network architectures for each task, as shown in Table 6.2. For example, on “Acrobot”, $|S| = 6$, $|A| = 1$, the policy network has hidden layers with 5 and 3 neurons, and the weight

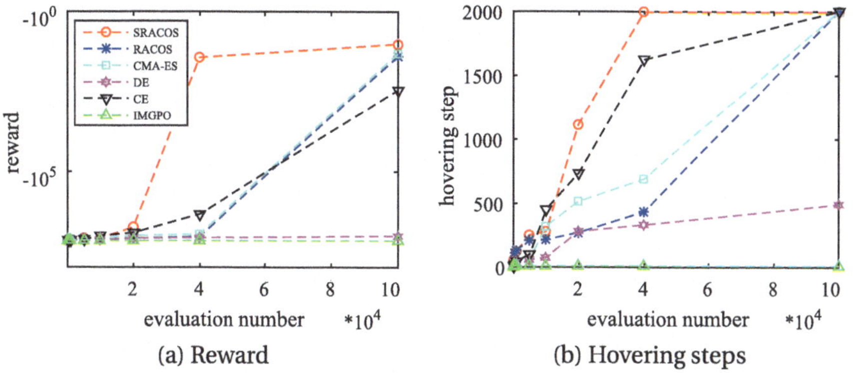

Fig. 6.5 The convergence speed on **a** the reward and **b** the hovering steps [4]

Table 6.1 Average performance of the reward, the hovering steps, and the success rate in the helicopter hovering control task [4]. The values in bold represent the best result in each item

Algorithms	Reward	Hovering step	Success rate
SRACOS	$\mathbf{-9.72 \times 10^5 \pm 2.17 \times 10^6}$	**1,837±364**	**4/15**
RACOS	$-3.18 \times 10^6 \pm 3.34 \times 10^6$	1,477±535	2/15
CMA-ES	$-5.29 \times 10^6 \pm 4.88 \times 10^6$	1,280±673	2/15
DE	$-1.02 \times 10^7 \pm 5.92 \times 10^5$	453±74	0/15
CE	$-5.48 \times 10^6 \pm 3.35 \times 10^6$	1,121±525	1/15
IMGPO	$-1.18 \times 10^7 \pm 2.66 \times 10^5$	256±31	0/15

Table 6.2 Parameters of the Gym tasks, including the dimensionality of the state space d_{State}, the number of actions, the layers and nodes of the feed-forward neural networks, the number of weights, and the horizon steps [4]

Task name	d_{State}	#Actions	NN nodes	#Weights	Horizon
Acrobot	6	1	5, 3	48	2,000
MountainCar	2	1	5	15	10,000
HalfCheetah	17	6	10	230	10,000
Humanoid	376	17	25	9825	50,000
Swimmer	8	2	5, 3	61	10,000
Ant	111	8	15	1785	10,000
Hopper	11	3	9, 5	159	10,000
LunarLander	8	1	5, 3	58	10,000

Table 6.3 The average reward and the standard deviation of the best found policy by each algorithm [4]. The numbers in bold represent the best cumulated reward in each row. The mark ↓ means the reward is better when smaller, and ↑ means better when larger

Method/Task	Acrobat↓	MountainCar↓	HalfCheetah↑	Humanoid↑
SRACOS	**156.60**±18.48	**132.40**±39.60	**36719.90**±8288.84	**502.57**±88.03
RACOS	169.70±14.15	141.50±0.97	27961.18±7493.08	398.03±19.23
CMA-ES	181.10±42.66	190.60±26.89	20191.83±984.95	357.09±124.77
DE	161.10±45.91	153.00±48.44	17250.21±305.01	428.97±67.89
CE	534.00±774.69	3048.90±4796.70	14714.05±5169.94	423.58±27.88
IMGPO	1545.00±736.14	5171.40±5090.29	10355.83±93.16	209.75±3.16
Method/Task	Swimmer↑	Ant↑	Hopper↑	LunarLander↑
SRACOS	**3692.65**±7.89	**2114.14**±501.11	**10818.98**±501.11	**238.14**±15.61
RACOS	3495.16±72.75	1215.28±1487.81	9892.70±417.85	193.45±35.62
CMA-ES	3202.33±11.98	63.66±12.00	9986.81±0.96	132.62±35.18
DE	3096.44±20.08	653.56±969.84	9931.70±1.35	125.00±93.86
CE	3002.26±46.14	722.88±531.73	5149.48±5006.35	92.45±110.81
IMGPO	270.73±3.27	42.52±3.57	136.28±23.04	64.29±27.32

dimension of the network is $|\boldsymbol{w}| = 48$. The maximum number of horizon steps is 2000. We run experiments for 15 times independently and report the test results of the best policy obtained by each algorithm in Table 6.3. It can be observed that SRACOS obtains the best results on all of these tasks. Especially on complex tasks from HalfCheetah to LunarLander, SRACOS drastically improves the average reward.

6.4 Summary

In this chapter, we improved the existing batch-mode derivative-free optimization framework and presented the sequential-mode framework, SRACOS, originally proposed in [3]. The sequential-mode derivative-free optimization can use the sample and its evaluation value at every step immediately, accelerating the updating process of optimization and improving the optimization efficiency. We also analyzed the query complexity of SRACOS and revealed the possibility that the sequential-mode optimization can be better than the batch-mode optimization from a theoretical perspective. The empirical results also demonstrated that SRACOS has a better convergence rate and scalability than the batch-mode RACOS algorithm. We also noticed the latest study proposed the RACE-CARS method using region-shrinking idea that achieves state-of-the-art sequential mode performance [1].

References

1. Han T, Li J, Guo Z, Jin Y (2025) Scalable acceleration for classification-based derivative-free optimization. In: Proceedings of the 39st AAAI conference on artificial intelligence (AAAI'25), Philadelphia, PA
2. Hansen N, Müller SD, Koumoutsakos P (2003) Reducing the time complexity of the derandomized evolution strategy with covariance matrix adaptation (CMA-ES). Evol Comput 11(1):1–18
3. Hu YQ, Qian H, Yu Y (2017) Sequential classification-based optimization for direct policy search. In: Proceedings of the 31st AAAI conference on artificial intelligence, pp 2029–2035, San Francisco, CA
4. Hu YQ, Qian H, Yu Y (2017) Sequential classification-based optimization for direct policy search. In: Proceedings of the 31st AAAI conference on artificial intelligence (AAAI'17), pp 2029–2035, San Francisco, CA
5. Kawaguchi K, Kaelbling LP, Lozano-Pérez T (2015) Bayesian optimization with exponential convergence. In: Advances in neural information processing systems, vol 28, pp7 91–2799, Montreal, Canada
6. Kim H, Jordan M, Sastry S, Ng A (2003) Autonomous helicopter flight via reinforcement learning. In: Advances in neural information processing systems
7. Rogier K, Whiteson S (2011) Neuroevolutionary reinforcement learning for generalized control of simulated helicopters. Evol Intell 4(4):219–241
8. Storn R, Price K (1997) Differential evolution-a simple and efficient heuristic for global optimization over continuous spaces. J Global Optim 11(4):341–359

Chapter 7
Optimization in High-Dimensional Search Space

Abstract This chapter addresses the challenge of optimizing high-dimensional functions, where traditional derivative-free optimization (DFO) methods struggle with scalability due to slow convergence and high computational costs. The focus is on problems with low optimal-effective dimensions, where only a small subspace significantly impacts the function value. The chapter introduces the Sequential Random Embeddings (SRE) technique, which sequentially applies random embeddings and employs DFO algorithms in each subspace to refine solutions. SRE reduces the embedding gap and improves optimization quality for a broad class of problems. The chapter is structured as follows: Sect. 7.1 defines functions with low effective dimensions, Sect. 7.2 discusses random embedding techniques, Sect. 7.3 introduces SRE, and Sect. 7.4 presents empirical studies on synthetic functions and classification tasks using the non-convex Ramp loss. Experimental results demonstrate that SRE significantly enhances the performance of state-of-the-art DFO methods in high-dimensional spaces, even for problems with up to 100,000 variables.

In the previous chapters, we introduced the sampling-and-learning (SAL) and sampling-and-classification (SAC) frameworks for derivative-free optimization (DFO) and proposed the RACOS algorithm based on these frameworks. While RACOS and other DFO methods have shown effectiveness in solving optimization problems typically with dimensionality smaller than 1,000, they often struggle with scalability when dealing with high-dimensional search spaces. This scalability issue can be attributed to the slow convergence rate in high dimensions, the high per-iteration computational cost, or both.

Existing studies have proposed two main directions to improve the scalability of DFO methods: decomposition and embedding. Decomposition methods extract subproblems from the original optimization problem and solve them to obtain a solution to the original problem. Embedding methods assume that the function value only depends on a small subspace of the high-dimensional space and optimize within that effective subspace. However, these approaches have limitations, such as relying on specific problem structures or assuming the existence of a clear effective subspace.

In this chapter, we study high-dimensional problems with low optimal ε-effective dimensions [6], where any variable can affect the function value, but only a small linear subspace has a significant impact, while the orthogonal complement subspace

Y. Yu et al., *Derivative-Free Optimization*, Machine Learning: Foundations, Methodologies, and Applications, https://doi.org/10.1007/978-981-96-5929-6_7

has a bounded small effect. We characterize the property of random embedding for such problems and present the sequential random embeddings (SRE) technique [6] to overcome the embedding gap. SRE applies random embedding multiple times sequentially and employs a DFO algorithm in each subspace to refine the solution. We also provide conditions under which SRE can improve the optimization quality for a broad class of problems.

Through extensive experiments on synthetic functions and classification tasks using the non-convex Ramp loss, we demonstrate that SRE can significantly enhance the performance of state-of-the-art DFO methods in high-dimensional problems, even for search spaces with up to 100,000 variables.

7.1 Functions with Low Effective Dimension

We first introduce the concept of Effective Dimension, which characterizes problems where the function value is affected by only a few effective dimensions [7, 8].

Definition 7.1 (*Effective Dimension*) A function $f : \mathbb{R}^D \to \mathbb{R}$ is said to have effective dimension d_e, with $d_e < D$, if there exists a linear subspace $\mathcal{V} \subseteq \mathbb{R}^D$ with dimension d_e such that for all $\boldsymbol{x} \in \mathbb{R}^D$,

$$f(\boldsymbol{x}) = f(\boldsymbol{x}_e + \boldsymbol{x}_c) = f(\boldsymbol{x}_e), \tag{7.1}$$

where $\boldsymbol{x}_e \in \mathcal{V}$, $\boldsymbol{x}_c \in \mathcal{V}^\perp$, and $\mathcal{V}^\perp$ denotes the orthogonal complement of $\mathcal{V}$. We call $\mathcal{V}$ the effective subspace of f and $\mathcal{V}^\perp$ the constant subspace.

Intuitively, Definition 7.1 means that f only varies along the effective subspace $\mathcal{V}$, while remaining constant along the orthogonal complement $\mathcal{V}^\perp$.

7.1.1 Random Embedding for Low Effective Dimension Problems

Random embedding is a technique that allows DFO algorithms to operate in a low-dimensional subspace of the high-dimensional search space [7, 8]. Given a high-dimensional function f and a random matrix $\boldsymbol{A} \in \mathbb{R}^{D\times d}$ with independent entries sampled from a Gaussian distribution $\mathcal{N}(0, \sigma^2)$, we construct a new optimization problem:

$$\min_{\boldsymbol{y}\in\mathbb{R}^d} g(\boldsymbol{y}) = f(\boldsymbol{A}\boldsymbol{y}), \tag{7.2}$$

where the solution space for g has dimension d. Each solution $\boldsymbol{y}$ is evaluated by mapping it back to the original high-dimensional space using $\boldsymbol{A}\boldsymbol{y}$.

The following lemma characterizes the effectiveness of random embedding for functions with low effective dimension [7, 8].

Lemma 7.1 *Given a function $f : \mathbb{R}^D \to \mathbb{R}$ with effective dimension d_e, and a random matrix $\boldsymbol{A} \in \mathbb{R}^{D\times d}$ with $d \geq d_e$ and independent entries sampled from $\mathcal{N}(0, \sigma^2)$, then, with probability 1, for any $\boldsymbol{x} \in \mathbb{R}^D$, there exists $\boldsymbol{y} \in \mathbb{R}^d$ such that $f(\boldsymbol{x}) = f(\boldsymbol{A}\boldsymbol{y})$.*

Proof Since f has effective dimension d_e, there exists an effective subspace $\mathcal{V} \subseteq \mathbb{R}^D$ with $\dim(\mathcal{V}) = d_e$. Any $\boldsymbol{x} \in \mathbb{R}^D$ can be decomposed as $\boldsymbol{x} = \boldsymbol{x}_e + \boldsymbol{x}_c$, where $\boldsymbol{x}_e \in \mathcal{V}$ and $\boldsymbol{x}_c \in \mathcal{V}^\perp$. By definition, $f(\boldsymbol{x}) = f(\boldsymbol{x}_e)$ for all $\boldsymbol{x}_e \in \mathcal{V}$. Thus, it suffices to show that, for any $\boldsymbol{x}_e \in \mathcal{V}$, there exists $\boldsymbol{y} \in \mathbb{R}^d$ such that $\boldsymbol{A}\boldsymbol{y} = \boldsymbol{x}_e$.

Let $\boldsymbol{\Phi} \in \mathbb{R}^{D\times d_e}$ be a matrix whose columns form a standard orthonormal basis for $\mathcal{V}$. For any $\boldsymbol{x}_e \in \mathcal{V}$, there exists $\boldsymbol{c} \in \mathbb{R}^{d_e}$ such that $\boldsymbol{x}_e = \boldsymbol{\Phi}\boldsymbol{c}$. Assuming that $\boldsymbol{\Phi}^\top \boldsymbol{A}$ has rank d_e (which will be proven later), there must exist $\boldsymbol{y} \in \mathbb{R}^d$ such that $\boldsymbol{\Phi}^\top \boldsymbol{A}\boldsymbol{y} = \boldsymbol{c}$, because $\text{rank}(\boldsymbol{\Phi}^\top \boldsymbol{A}) = \text{rank}([\boldsymbol{\Phi}^\top \boldsymbol{A}, \boldsymbol{c}])$. Multiplying both sides by $\boldsymbol{\Phi}$, we have $\boldsymbol{A}\boldsymbol{y} = \boldsymbol{\Phi}\boldsymbol{\Phi}^\top \boldsymbol{A}\boldsymbol{y} = \boldsymbol{\Phi}\boldsymbol{c} = \boldsymbol{x}_e$.

It remains to prove that $\boldsymbol{\Phi}^\top \boldsymbol{A}$ has rank d_e with probability 1. Let $\boldsymbol{A}_e \in \mathbb{R}^{D\times d_e}$ be a submatrix of $\boldsymbol{A}$ consisting of any d_e columns of $\boldsymbol{A}$, which are i.i.d. samples from $\mathcal{N}(\boldsymbol{0}, \sigma^2 \boldsymbol{I})$. By the orthonormality of $\boldsymbol{\Phi}$, the columns of $\boldsymbol{\Phi}^\top \boldsymbol{A}_e$ are i.i.d. samples from $\mathcal{N}(\boldsymbol{0}, \sigma^2 \boldsymbol{I}_{d_e})$. The set of singular matrices in $\mathbb{R}^{d_e\times d_e}$ has Lebesgue measure zero, and the Gaussian distribution is absolutely continuous with respect to the Lebesgue measure. Therefore, $\boldsymbol{\Phi}^\top \boldsymbol{A}_e$ is almost surely non-singular, and the same holds for $\boldsymbol{\Phi}^\top \boldsymbol{A}$. □

Lemma 7.1 implies that, given a random embedding matrix $\boldsymbol{A} \in \mathbb{R}^{D\times d}$, for any minimizer $\boldsymbol{x}^* \in \mathbb{R}^D$ of f, there exists $\boldsymbol{y}^* \in \mathbb{R}^d$ such that $f(\boldsymbol{A}\boldsymbol{y}^*) = f(\boldsymbol{x}^*)$. Thus, we can optimize the lower dimensional function $g(\boldsymbol{y}) = f(\boldsymbol{A}\boldsymbol{y})$ instead of the original high-dimensional $f(\boldsymbol{x})$.

7.2 Optimal ε-Effective Dimension

The Effective Dimension assumption in Definition 7.1 requires the existence of a linear subspace that has exactly zero effect on the function value. This assumption may be too strict for real-world problems. We relax this assumption by introducing the concept of optimal ε-effective dimension.

Definition 7.2 (*Optimal ε-Effective Dimension*) For any $\varepsilon > 0$, a function $f : \mathbb{R}^D \to \mathbb{R}$ is said to have an ε-effective subspace $\mathcal{V}_\varepsilon$ if there exists a linear subspace $\mathcal{V}_\varepsilon \subseteq \mathbb{R}^D$ such that for all $\boldsymbol{x} \in \mathbb{R}^D$,

$$|f(\boldsymbol{x}) - f(\boldsymbol{x}_\varepsilon)| \leq \varepsilon, \tag{7.3}$$

where $\boldsymbol{x}_\varepsilon \in \mathcal{V}_\varepsilon$ is the orthogonal projection of $\boldsymbol{x}$ onto $\mathcal{V}_\varepsilon$. Let $\mathbb{V}_\varepsilon$ denote the collection of all ε-effective subspaces of f. The optimal ε-effective dimension of f is defined as

$$d_\varepsilon = \min_{\mathcal{V}_\varepsilon \in \mathbb{V}_\varepsilon} \dim(\mathcal{V}_\varepsilon), \tag{7.4}$$

where $\dim(\mathcal{V})$ denotes the dimension of a linear subspace $\mathcal{V}$.

Note that ε and d_ε are related variables: a small d_ε often implies a large ε, while a small ε implies a large d_ε.

7.2.1 Random Embedding for Problems with Low Optimal ε-Effective Dimension

The following lemma characterizes the effect of random embedding for functions with low optimal ε-effective dimension.

Lemma 7.2 *Given a function $f : \mathbb{R}^D \to \mathbb{R}$ with optimal ε-effective dimension d_ε, and a random matrix $\boldsymbol{A} \in \mathbb{R}^{D\times d}$ with $d \geq d_\varepsilon$ and independent entries sampled from $\mathcal{N}(0, \sigma^2)$, then, with probability 1, for any $\boldsymbol{x} \in \mathbb{R}^D$, there exists $\boldsymbol{y} \in \mathbb{R}^d$ such that $|f(\boldsymbol{x}) - f(\boldsymbol{A}\boldsymbol{y})| \leq 2\varepsilon$.*

Proof The proof follows a similar argument as in Lemma 7.1. Since f has optimal ε-effective dimension d_ε, there exists an ε-effective subspace $\mathcal{V}_\varepsilon \subseteq \mathbb{R}^D$ with $\dim(\mathcal{V}_\varepsilon) = d_\varepsilon$. Any $\boldsymbol{x} \in \mathbb{R}^D$ can be decomposed as $\boldsymbol{x} = \boldsymbol{x}_\varepsilon + \boldsymbol{x}_\varepsilon^\perp$, where $\boldsymbol{x}_\varepsilon \in \mathcal{V}_\varepsilon$ and $\boldsymbol{x}_\varepsilon^\perp \in \mathcal{V}_\varepsilon^\perp$. By definition, $|f(\boldsymbol{x}) - f(\boldsymbol{x}_\varepsilon)| \leq \varepsilon$. Thus, it suffices to show that, for any $\boldsymbol{x}_\varepsilon \in \mathcal{V}_\varepsilon$, there exists $\boldsymbol{y} \in \mathbb{R}^d$ such that $|f(\boldsymbol{x}_\varepsilon) - f(\boldsymbol{A}\boldsymbol{y})| \leq \varepsilon$.

Following the same steps as in Lemma 7.1, we can prove that there exists $\boldsymbol{y} \in \mathbb{R}^d$ such that $\boldsymbol{A}\boldsymbol{y} = \boldsymbol{x}_\varepsilon + \tilde{\boldsymbol{x}}$, where $\tilde{\boldsymbol{x}} \in \mathcal{V}_\varepsilon^\perp$. Since $\boldsymbol{A}\boldsymbol{y} \in \mathbb{R}^D$, by the definition of ε-effective subspace, we have $|f(\boldsymbol{x}_\varepsilon) - f(\boldsymbol{A}\boldsymbol{y})| \leq \varepsilon$. Combining this with $|f(\boldsymbol{x}) - f(\boldsymbol{x}_\varepsilon)| \leq \varepsilon$, we conclude that $|f(\boldsymbol{x}) - f(\boldsymbol{A}\boldsymbol{y})| \leq 2\varepsilon$. □

Lemma 7.2 implies that, given a random embedding matrix $\boldsymbol{A} \in \mathbb{R}^{D\times d}$, for any minimizer $\boldsymbol{x}^* \in \mathbb{R}^D$ of f, there exists $\tilde{\boldsymbol{y}} \in \mathbb{R}^d$ such that $f(\boldsymbol{A}\tilde{\boldsymbol{y}}) - f(\boldsymbol{x}^*) \leq 2\varepsilon$. This embedding gap grows twice as fast as ε.

7.2.2 Optimization with Random Embedding

Given a high-dimensional function f and a random matrix $\boldsymbol{A} \in \mathbb{R}^{D\times d}$, we construct a new optimization problem:

$$\min_{\boldsymbol{y}\in\mathbb{R}^d} g(\boldsymbol{y}) = f(\boldsymbol{A}\boldsymbol{y}), \tag{7.5}$$

where the solution space for g has dimension d. Each solution $\boldsymbol{y}$ is evaluated by mapping it back to the original high-dimensional space using $\boldsymbol{A}\boldsymbol{y}$.

For functions with low optimal ε-effective dimension, we can bound the gap between the optimal function values of g and f based on Lemma 7.2.

Theorem 7.1 *Given a function $f : \mathbb{R}^D \rightarrow \mathbb{R}$ with optimal ε-effective dimension d_ε, and a random matrix $\boldsymbol{A} \in \mathbb{R}^{D\times d}$ with $d \geq d_\varepsilon$ and independent entries sampled from $\mathcal{N}(0, \sigma^2)$, let $\boldsymbol{x}^*$ be a global minimizer of f. Then, with probability 1,*

$$\min_{\boldsymbol{y}\in\mathbb{R}^d} g(\boldsymbol{y}) - f(\boldsymbol{x}^*) \leq 2\varepsilon. \tag{7.6}$$

Proof The proof follows directly from Lemma 7.2. □

Let $\tilde{\boldsymbol{y}}$ be the solution found by a DFO algorithm in the low-dimensional space. There is an approximation gap between $g(\tilde{\boldsymbol{y}})$ and $\min_{\boldsymbol{y}\in\mathbb{R}^d} g(\boldsymbol{y})$, which depends on the dimension d, the function complexity, and the optimization budget. We assume that this approximation gap is upper bounded by θ. Furthermore, as shown in Theorem 7.1, there exists an embedding gap of 2ε, which cannot be compensated by the optimization algorithm. Thus, the simple regret of the algorithm is upper bounded by the sum of the approximation gap and the embedding gap:

$$g(\tilde{\boldsymbol{y}}) - f(\boldsymbol{x}^*) \leq \theta + 2\varepsilon. \tag{7.7}$$

7.3 Sequential Random Embeddings

To reduce the embedding gap while keeping the approximation gap unaffected, we present the sequential random embeddings (SRE) technique. SRE applies random embedding multiple times sequentially and employs a DFO algorithm in each subspace to refine the solution.

Let $\tilde{\boldsymbol{x}}_1 = \boldsymbol{0}$ and $\mathcal{S}_i = \{\boldsymbol{A}^{(i)}\boldsymbol{y} \mid \boldsymbol{y} \in \mathbb{R}^d\}$ denote the subspace defined by the random matrix $\boldsymbol{A}^{(i)}$, where $i = 1, \ldots, m$. The SRE procedure can be described as follows:

- In the first step, generate a random matrix $\boldsymbol{A}^{(1)}$ defining a subspace $\mathcal{S}_1$, and apply a DFO algorithm to find a near-optimal solution in the subspace: $\tilde{\boldsymbol{y}}_1 = \arg\min_{\boldsymbol{y}} f(\boldsymbol{A}^{(1)}\boldsymbol{y})$. Let $\tilde{\boldsymbol{x}}_2 = \boldsymbol{A}^{(1)}\tilde{\boldsymbol{y}}_1$ be the high-dimensional solution.
- In the second step, generate another random matrix $\boldsymbol{A}^{(2)}$ defining a subspace $\mathcal{S}_2$, and apply the DFO algorithm to optimize the residue of the current solution $\tilde{\boldsymbol{x}}_2$ in the subspace: $\tilde{\boldsymbol{y}}_2 = \arg\min_{\boldsymbol{y}} f(\tilde{\boldsymbol{x}}_2 + \boldsymbol{A}^{(2)}\boldsymbol{y})$. Update the current solution $\tilde{\boldsymbol{x}}_3 = \tilde{\boldsymbol{x}}_2 + \boldsymbol{A}^{(2)}\tilde{\boldsymbol{y}}_2$.
- In the following steps, repeat the process of optimizing the residue in each subspace.

Let $\boldsymbol{x}^* - \tilde{\boldsymbol{x}}_i$ be the residue solution to be approximated in the ith step of SRE, and let $\hat{\boldsymbol{x}}_i$ be the orthogonal projection of $\boldsymbol{x}^* - \tilde{\boldsymbol{x}}_i$ onto the subspace $\mathcal{S}_i$. We define the embedding ratio as $\|\hat{\boldsymbol{x}}_i\|/\|\boldsymbol{x}^* - \tilde{\boldsymbol{x}}_i\|$ and the optimization ratio as $\|\hat{\boldsymbol{x}}_i - \boldsymbol{A}^{(i)}\tilde{\boldsymbol{y}}_i\|/\|\hat{\boldsymbol{x}}_i\|$.

The following theorem provides a condition under which SRE can strictly reduce the solution gap in each step.

Theorem 7.2 *Given a function $f : \mathbb{R}^D \to \mathbb{R}$ with optimal ε-effective dimension d_ε, and a sequence of random matrices $\{\boldsymbol{A}^{(i)}\}_{i=1}^m \subseteq \mathbb{R}^{D\times d}$ with $d \geq d_\varepsilon$ and independent entries sampled from $\mathcal{N}(0, \sigma^2)$, let $\boldsymbol{x}^*$ be a global minimizer of f. For all $i = 1, \ldots, m$, if*

$$\frac{\|\hat{\boldsymbol{x}}_i - \boldsymbol{A}^{(i)}\tilde{\boldsymbol{y}}_i\|}{\|\hat{\boldsymbol{x}}_i\|} \leq \frac{1}{5}\cdot\frac{\|\hat{\boldsymbol{x}}_i\|}{\|\boldsymbol{x}^* - \tilde{\boldsymbol{x}}_i\|}, \tag{7.8}$$

then $\|\boldsymbol{x}^* - \tilde{\boldsymbol{x}}_i\| > \|\boldsymbol{x}^* - \tilde{\boldsymbol{x}}_{i+1}\|$.

Proof The proof follows a similar argument as in [11]. For any $i = 1, \ldots, m$, since $\hat{\boldsymbol{x}}_i$ is the orthogonal projection of $\boldsymbol{x}^* - \tilde{\boldsymbol{x}}_i$ onto $\mathcal{S}_i$, we have

$$\begin{aligned}
\|\boldsymbol{x}^* - \tilde{\boldsymbol{x}}_i\|^2 &= \|\boldsymbol{x}^* - \tilde{\boldsymbol{x}}_i - \hat{\boldsymbol{x}}_i\|^2 + \|\hat{\boldsymbol{x}}_i\|^2\\
&\geq (\|\boldsymbol{x}^* - \tilde{\boldsymbol{x}}_i - \boldsymbol{A}^{(i)}\tilde{\boldsymbol{y}}_i\| - \|\hat{\boldsymbol{x}}_i - \boldsymbol{A}^{(i)}\tilde{\boldsymbol{y}}_i\|)^2 + \|\hat{\boldsymbol{x}}_i\|^2\\
&= \|\boldsymbol{x}^* - \tilde{\boldsymbol{x}}_{i+1}\|^2 + \|\hat{\boldsymbol{x}}_i - \boldsymbol{A}^{(i)}\tilde{\boldsymbol{y}}_i\|^2 + \|\hat{\boldsymbol{x}}_i\|^2\\
&\quad - 2\|\boldsymbol{x}^* - \tilde{\boldsymbol{x}}_i - \boldsymbol{A}^{(i)}\tilde{\boldsymbol{y}}_i\|\cdot\|\hat{\boldsymbol{x}}_i - \boldsymbol{A}^{(i)}\tilde{\boldsymbol{y}}_i\|\\
&\geq \|\boldsymbol{x}^* - \tilde{\boldsymbol{x}}_{i+1}\|^2 + (\|\hat{\boldsymbol{x}}_i - \boldsymbol{A}^{(i)}\tilde{\boldsymbol{y}}_i\| - \|\hat{\boldsymbol{x}}_i\|)^2\\
&\quad - 2(\|\boldsymbol{x}^* - \tilde{\boldsymbol{x}}_i\| + \|\boldsymbol{A}^{(i)}\tilde{\boldsymbol{y}}_i\|)\cdot\|\hat{\boldsymbol{x}}_i - \boldsymbol{A}^{(i)}\tilde{\boldsymbol{y}}_i\|\\
&\quad + 2\|\hat{\boldsymbol{x}}_i\|\cdot\|\hat{\boldsymbol{x}}_i - \boldsymbol{A}^{(i)}\tilde{\boldsymbol{y}}_i\|\\
&\geq \|\boldsymbol{x}^* - \tilde{\boldsymbol{x}}_{i+1}\|^2 + (\|\hat{\boldsymbol{x}}_i - \boldsymbol{A}^{(i)}\tilde{\boldsymbol{y}}_i\| - \|\hat{\boldsymbol{x}}_i\|)^2\\
&\quad - 2\|\boldsymbol{x}^* - \tilde{\boldsymbol{x}}_i\|\cdot\|\hat{\boldsymbol{x}}_i - \boldsymbol{A}^{(i)}\tilde{\boldsymbol{y}}_i\| - 2\|\hat{\boldsymbol{x}}_i - \boldsymbol{A}^{(i)}\tilde{\boldsymbol{y}}_i\|^2,
\end{aligned}$$

where the last inequality follows from $\|\boldsymbol{A}^{(i)}\tilde{\boldsymbol{y}}_i\| - \|\hat{\boldsymbol{x}}_i\| \leq \|\hat{\boldsymbol{x}}_i - \boldsymbol{A}^{(i)}\tilde{\boldsymbol{y}}_i\|$.

Since $\|\hat{\boldsymbol{x}}_i - \boldsymbol{A}^{(i)}\tilde{\boldsymbol{y}}_i\|\cdot\|\boldsymbol{x}^* - \tilde{\boldsymbol{x}}_i\|/\|\hat{\boldsymbol{x}}_i\|^2 \leq 1/5$ and $\|\hat{\boldsymbol{x}}_i\| \leq \|\boldsymbol{x}^* - \tilde{\boldsymbol{x}}_i\|$, we have

$$(\|\hat{\boldsymbol{x}}_i - \boldsymbol{A}^{(i)}\tilde{\boldsymbol{y}}_i\| - \|\hat{\boldsymbol{x}}_i\|)^2 - 2\|\boldsymbol{x}^* - \tilde{\boldsymbol{x}}_i\|\cdot\|\hat{\boldsymbol{x}}_i - \boldsymbol{A}^{(i)}\tilde{\boldsymbol{y}}_i\| - 2\|\hat{\boldsymbol{x}}_i - \boldsymbol{A}^{(i)}\tilde{\boldsymbol{y}}_i\|^2 > 0.$$

Therefore, $\|\boldsymbol{x}^* - \tilde{\boldsymbol{x}}_i\| > \|\boldsymbol{x}^* - \tilde{\boldsymbol{x}}_{i+1}\|$ for all $i = 1, \ldots, m$. □

Theorem 7.2 suggests that, under a mild condition on the optimization ratio, SRE can reduce the solution gap in each step for a broad class of problems with local Holder continuity, defined as follows.

Definition 7.3 (*Local Holder Continuity*) A function $f : \mathbb{R}^D \to \mathbb{R}$ satisfies local Holder continuity if there exist constants $L, \alpha > 0$ such that, for all $\boldsymbol{x} \in \mathbb{R}^D$,

$$f(\boldsymbol{x}) - f(\boldsymbol{x}^*) \leq L\cdot\|\boldsymbol{x} - \boldsymbol{x}^*\|_2^\alpha, \tag{7.9}$$

where $\boldsymbol{x}^*$ is a global minimizer of f.

Local Holder continuity allows the function to have many local optima or be non-differentiable, as long as the rate of increase around the global minimizer is bounded.

7.3.1 Less Greedy SRE

In the SRE procedure described above, each sub-problem in the subspace is solved greedily. However, a perfect solution for one sub-problem may not be globally optimal, and once an unsatisfactory solution is found, it is difficult to correct it in later steps due to the greedy process. To address this issue, we introduce a withdrawal variable β to the previous solution, allowing the algorithm to eliminate the previous solution if necessary. The optimization problem in each step becomes

$$\min_{\boldsymbol{y},\beta} f(\beta\tilde{\boldsymbol{x}}_i + \boldsymbol{A}^{(i)}\boldsymbol{y}). \tag{7.10}$$

Since DFO methods make few assumptions about the optimization problem, we can simply let the algorithm optimize β together with $\boldsymbol{y}$.

The full SRE algorithm is presented in Algorithm 7.1.

Algorithm 7.1 Sequential Random Embeddings (SRE)

Require:
Objective function f;
DFO algorithm $\mathcal{M}$;
Number of function evaluations n;
Upper bound of optimal ε-effective dimension d;
Number of sequential random embeddings m.
Ensure:
1: $\tilde{\boldsymbol{x}}_1 = \boldsymbol{0}$.
2: **for** $i = 1$ to m **do**
3: Sample a random matrix $\boldsymbol{A}^{(i)} \in \mathbb{R}^{D\times d}$ with entries from $\mathcal{N}(0, 1/d)$.
4: Apply $\mathcal{M}$ to optimize $g_i(\boldsymbol{y}) = f(\beta\tilde{\boldsymbol{x}}_i + \boldsymbol{A}^{(i)}\boldsymbol{y})$ with n/m function evaluations.
5: Obtain the solution $\tilde{\boldsymbol{y}}_i$ and β_i for $g_i(\boldsymbol{y})$ using $\mathcal{M}$.
6: $\tilde{\boldsymbol{x}}_{i+1} = \beta_i\tilde{\boldsymbol{x}}_i + \boldsymbol{A}^{(i)}\tilde{\boldsymbol{y}}_i$.
7: **end for**
8: **return** $\arg\min_{i=2,\ldots,m+1} f(\tilde{\boldsymbol{x}}_i)$.

7.4 Empirical Study

We empirically evaluate the effectiveness of SRE in combination with state-of-the-art DFO methods on synthetic functions and classification tasks using the non-convex Ramp loss.

7.4.1 Experimental Setup

We consider the high-dimensional search space $\mathcal{X} = [-u, u]^D$ and the low-dimensional search space $\mathcal{Y} = [-l, l]^d$, where $u, l > 0$. The random matrix $\boldsymbol{A} \in \mathbb{R}^{D \times d}$ has independent entries sampled from $\mathcal{N}(0, 1/d)$.

To handle the case where $\boldsymbol{A}\boldsymbol{y}' \notin \mathcal{X}$ for some $\boldsymbol{y}' \in \mathcal{Y}$, we employ Euclidean projection $P_{\mathcal{X}}(\boldsymbol{A}\boldsymbol{y}') = \arg\min_{\boldsymbol{x} \in \mathcal{X}} \|\boldsymbol{x} - \boldsymbol{A}\boldsymbol{y}'\|_2$. The function value of $\boldsymbol{A}\boldsymbol{y}'$ is then set to $f(P_{\mathcal{X}}(\boldsymbol{A}\boldsymbol{y}')) + \|P_{\mathcal{X}}(\boldsymbol{A}\boldsymbol{y}') - \boldsymbol{A}\boldsymbol{y}'\|_1$.

We apply SRE to three state-of-the-art DFO methods: IMGPO [4], CMAES [3], and RACOS [9]. The prefix "RE-" denotes the single random embedding variant, while "SRE-" denotes the sequential random embeddings variant. Random search is included as a baseline.

7.4.2 Synthetic Functions

We construct high-dimensional versions of the Sphere and Ackley functions that satisfy the optimal ε-effective dimension assumption. The high-dimensional Sphere function is defined as

$$f_1(\boldsymbol{x}) = \sum_{i=1}^{10} ([\boldsymbol{x}]_i - 0.2)^2 + \frac{1}{D} \sum_{i=11}^{D} ([\boldsymbol{x}]_i - 0.2)^2, \tag{7.11}$$

where $[\boldsymbol{x}]_i$ denotes the ith coordinate of $\boldsymbol{x}$. The high-dimensional Ackley function is defined as

$$f_2(\boldsymbol{x}) = -20 \exp\left(-\frac{1}{5}\sqrt{\frac{1}{10}\sum_{i=1}^{10} ([\boldsymbol{x}]_i - 0.2)^2}\right) \tag{7.12}$$

$$- \exp\left(\frac{1}{10}\sum_{i=1}^{10} \cos(2\pi([\boldsymbol{x}]_i - 0.2))\right) + e + 20 \tag{7.13}$$

$$+ \frac{1}{D}\sum_{i=11}^{D} ([\boldsymbol{x}]_i - 0.2)^2. \tag{7.14}$$

The optimal solution for both functions is $\boldsymbol{x}^* = (0.2, \ldots, 0.2)$. We set $\mathcal{X} = [-1, 1]^D$, $\mathcal{Y} = [-1, 1]^d$, and $\beta \in [-1, 1]$. Each algorithm is run 30 times independently, and the average performance is reported.

Effect of the Number of Random Embeddings

We investigate the effect of the number of random embeddings m in SRE. We set $D = 10000$, $n = 10000$ (total number of function evaluations), $d = 10$, and vary $m \in \{1, 2, 5, 8, 10, 20\}$. Note that when $m = 1$, SRE degenerates to RE.

Figure 7.1 shows that, for a fixed total number of function evaluations, there is a trade-off in choosing m. If m is too large, the budget for each step of SRE is limited, while if m is too small, the number of steps in SRE is limited. Both scenarios can lead to unsatisfactory optimization performance.

Effect of Subspace Dimension

We study how the low-dimensional subspace size d affects the optimization performance of SRE-based algorithms. We set $D = 10000$, $n = 10000$, $m = 5$, and vary $d \in \{1, 5, 8, 10, 12, 15, 20\}$.

Figure 7.2 demonstrates that, in most cases, the closer d is to the optimal ε-effective dimension d_ε, the better the optimization performance. This highlights the importance of having a good estimate of d_ε. Moreover, even when $d < d_\varepsilon = 10$ but close to d_ε, the performance of SRE-based algorithms remains satisfactory.

Scalability

We investigate the scalability of the algorithms with respect to the search space dimension D. We set $D \in \{100, 500, 1000, 5000, 10000\}$, $n = 10000$, $d = 10$ for RE and SRE, and $m = 5$ for SRE.

Figure 7.3 shows that SRE-based algorithms have the lowest growth rate as the dimension increases, while algorithms without RE have the highest growth rate. This indicates that SRE can effectively scale DFO algorithms to high-dimensional problems.

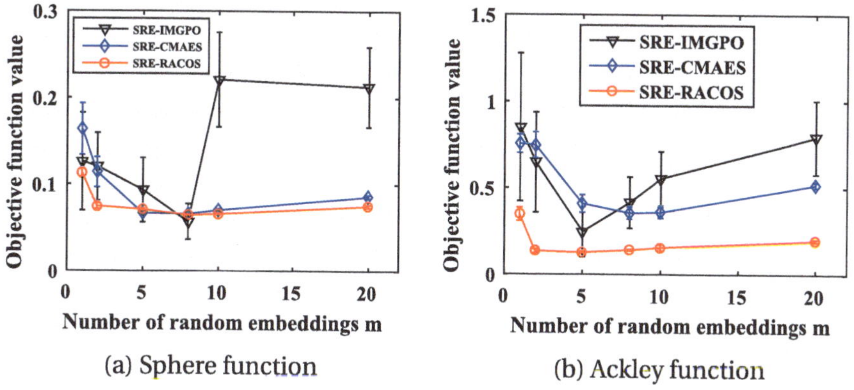

Fig. 7.1 Effect of the number of random embeddings m on the optimization performance [6]

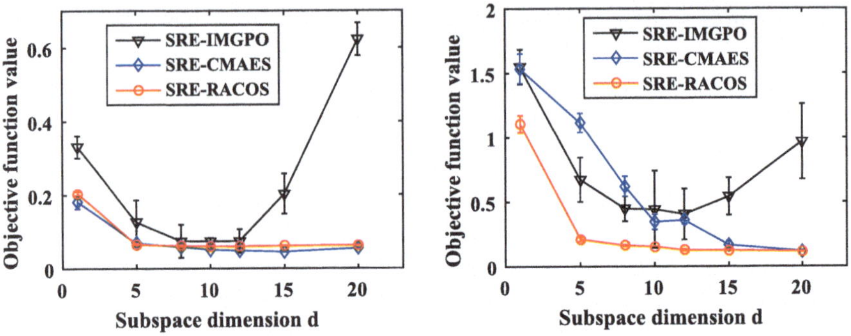

(a) Sphere function (b) Ackley function

Fig. 7.2 Effect of the subspace dimension d on the optimization performance [6]

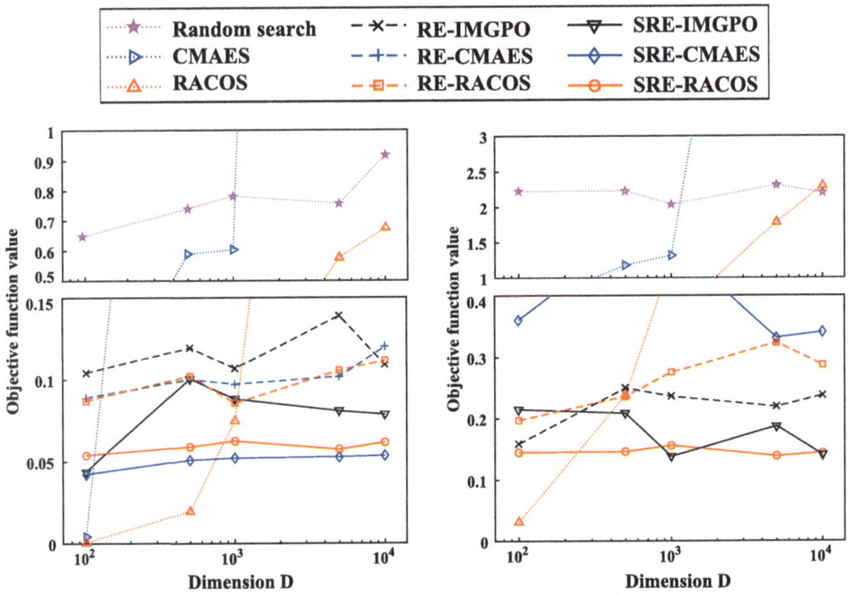

(a) Sphere function (b) Ackley function

Fig. 7.3 Scalability of the algorithms with respect to the search space dimension D [6]

Convergence Rate

We examine the convergence rate of the algorithms with respect to the number of function evaluations. We set $D = 10000$, $n \in \{2000, 4000, 6000, 8000, 10000\}$, $d = 10$ for RE and SRE, and $m = 5$ for SRE.

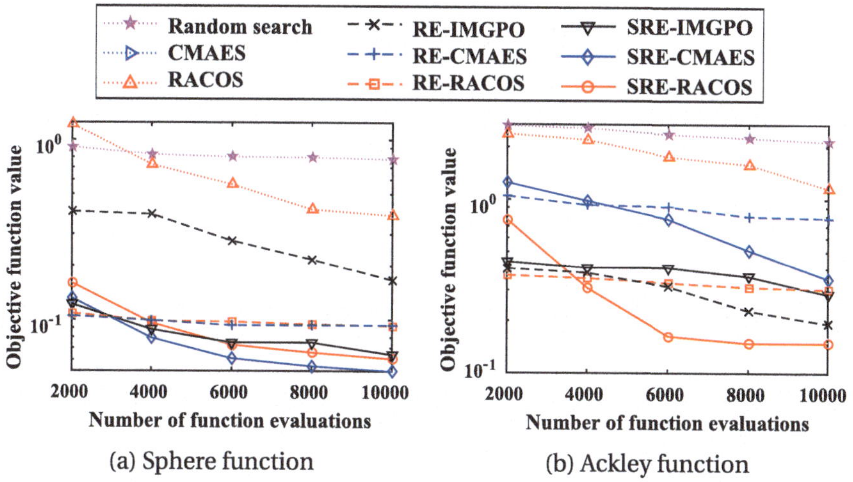

Fig. 7.4 Convergence rate of the algorithms with respect to the number of function evaluations [6]

Figure 7.4 shows that SRE-based algorithms generally reduce the objective function value at the highest rate, while algorithms without RE have the lowest convergence rate. This suggests that SRE can accelerate the convergence of DFO algorithms in high-dimensional problems.

7.4.3 Classification with Ramp Loss

We evaluate the algorithms on a classification task using the non-convex Ramp loss [2]. The Ramp loss is defined as $R_s(z) = H_1(z) - H_s(z)$ with $s < 1$, where $H_s(z) = \max\{0, s - z\}$ is the Hinge loss with hinge point s. The objective is to find a vector $\boldsymbol{w}$ and a scalar b that minimize:

$$f(\boldsymbol{w}, b) = \frac{1}{2}\|\boldsymbol{w}\|_2^2 + C\sum_{\ell=1}^{L} R_s(y_\ell(\boldsymbol{w}^\top \boldsymbol{v}_\ell + b)), \tag{7.15}$$

where $\boldsymbol{v}_\ell$ is the ℓth training instance and $y_\ell \in \{-1, +1\}$ is its corresponding label.

We employ four binary-class UCI datasets [1]: Gisette, Arcene, Dexter, and Dorothea, with feature dimensions D of 5×10^3, 10^4, 2×10^4, and 10^5, respectively.

To study the algorithms' effectiveness under different hyper-parameter settings, we test $s \in \{-1, 0\}$ and $C \in \{0.1, 0.5, 1, 2, 5, 10\}$. We set $d = 20$, $n = 3D$, $\mathcal{X} = [-10, 10]^D$, $\mathcal{Y} = [-10, 10]^d$, and $\beta \in [-10, 10]$ for all algorithms except CCCP [10], a gradient-based non-convex optimization method. For CCCP, we set $\mathcal{X} = [-10, 10]^D$ and let it run until convergence. For SRE-based algorithms, we set $m = 5$.

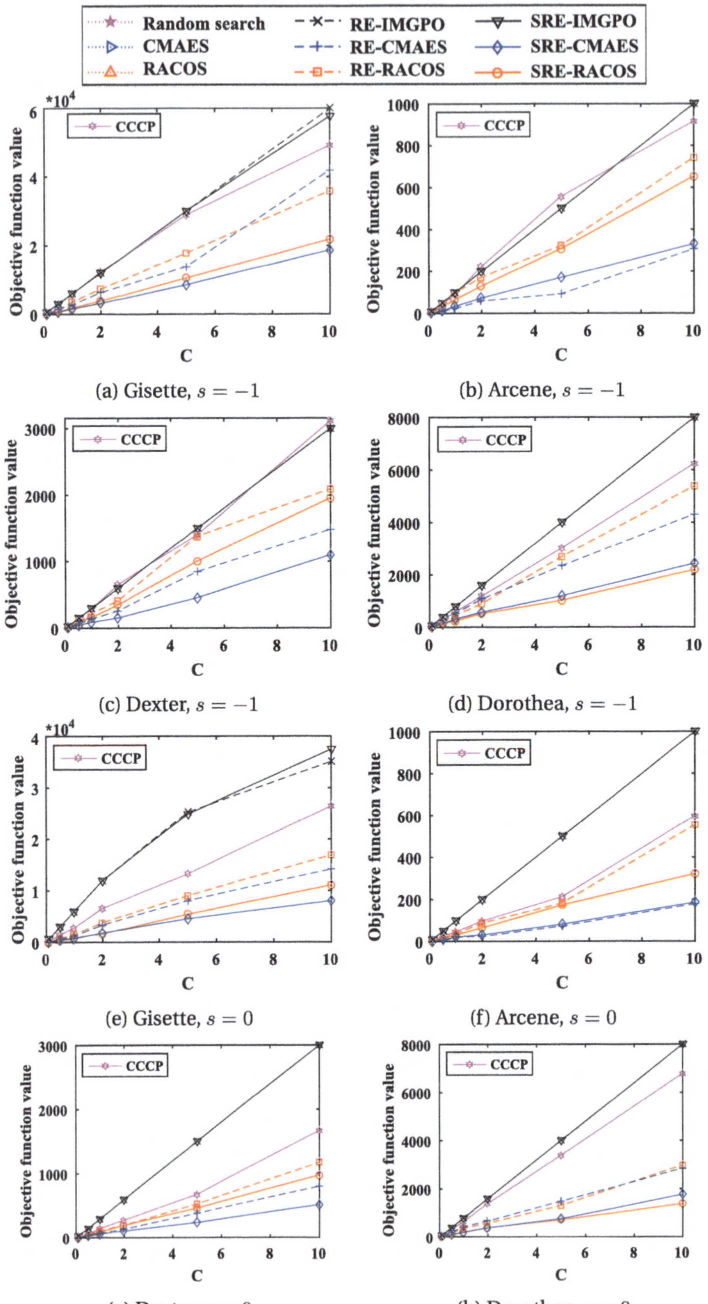

(a) Gisette, $s = -1$ (b) Arcene, $s = -1$

(c) Dexter, $s = -1$ (d) Dorothea, $s = -1$

(e) Gisette, $s = 0$ (f) Arcene, $s = 0$

(g) Dexter, $s = 0$ (h) Dorothea, $s = 0$

Fig. 7.5 Achieved objective function values of the algorithms on the classification task with Ramp loss [6]

Figure 7.5 reports the achieved objective function values on each dataset. SRE-based algorithms consistently outperform other methods, except on the Arcene dataset where RE-based algorithms achieve the best performance. This verifies the effectiveness of SRE and RE, with SRE being more effective than RE in most cases. Moreover, SRE-based algorithms significantly outperform CCCP in terms of optimization performance.

7.5 Summary

This chapter investigated high-dimensional optimization problems where all variables can affect the function value, but many of them have only a small bounded effect. We defined such problems as functions with a low optimal ε-effective dimension and showed that single random embedding incurs a 2ε loss that cannot be compensated by the subsequent optimization algorithm.

To address this issue, we presented the sequential random embeddings (SRE) technique [5], which applies random embedding multiple times sequentially and employs a DFO algorithm in each subspace to refine the solution. We provided conditions under which SRE can strictly reduce the embedding loss in each step for a broad class of problems.

Empirical results on synthetic functions and classification tasks using the non-convex Ramp loss demonstrated that SRE can significantly enhance the performance of state-of-the-art DFO methods in high-dimensional problems, even for search spaces with up to 100,000 variables. These findings highlight the potential of SRE in scaling DFO algorithms to tackle complex high-dimensional optimization problems.

References

1. Blake CL, Keogh E, Merz CJ (1998) UCI Repository of machine learning databases. http://www.ics.uci.edu/~mlearn/MLRepository.html
2. Collobert R, Sinz F, Weston J, Bottou L (2006) Trading convexity for scalability. In: Proceedings of the 23rd international conference on machine learning, Pittsburgh, Pennsylvania, pp 201–208
3. Hansen N, Müller SD, Koumoutsakos P (2003) Reducing the time complexity of the derandomized evolution strategy with covariance matrix adaptation (CMA-ES). Evol Comput 11(1):1–18
4. Kawaguchi K, Kaelbling LP, Lozano-Pérez T (2015) Bayesian optimization with exponential convergence. In: Advances in neural information processing systems, vol 28, Montreal, Canada, pp 2791–2799
5. Qian H, Hu YQ, Yu Y (2016) Derivative-free optimization of high-dimensional non-convex functions by sequential random embeddings. In: Proceedings of the 25th international joint conference on artificial intelligence, New York, NY, pp 1946–1952
6. Qian H, Hu YQ, Yu Y (2016) Derivative-free optimization of high-dimensional non-convex functions by sequential random embeddings. In: Proceedings of the 25th international joint conference on artificial intelligence, New York, NY, pp 1946–1952

7. Wang Z, Zoghi M, Hutter F, Matheson D, De Freitas N (2013) Bayesian optimization in high dimensions via random embeddings. In: Proceedings of the 23rd international joint conference on artificial intelligence, Beijing, China, pp 1778–1784
8. Wang Z, Hutter F, Zoghi M, Matheson D, de Freitas N (2016) Bayesian optimization in a billion dimensions via random embeddings. J Artif Intell Res 55:361–387
9. Yu Y, Qian H, Hu YQ (2016) Derivative-free optimization via classification. In: Proceedings of the 30th AAAI conference on artificial intelligence, Phoenix, Arizona, pp 2286–2292
10. Yuille AL, Rangarajan A (2001) The concave-convex procedure (CCCP). In: Advances in neural information processing systems, vol 14, Vancouver, Canada, pp 1033–1040
11. Zhang L, Mahdavi M, Jin R, Yang T, Zhu S (2013) Recovering the optimal solution by dual random projection. In: Proceedings of the 26th conference on learning theory, Princeton, NJ, pp 135–157

Chapter 8
Optimization Under Noise

Abstract This chapter addresses the challenge of optimizing noisy objective functions in derivative-free optimization (DFO), a common issue in real-world applications like reinforcement learning. While the Racos algorithm is effective in noise-free environments, it struggles with noisy evaluations. The chapter introduces value suppression, a novel noise-handling mechanism that delays noise mitigation until the best-so-far solution stagnates, reducing computational costs compared to traditional methods like sampling and threshold selection. The mechanism is integrated into the SRacos algorithm, resulting in SSRacos, which is shown to outperform other noise-handling techniques in both synthetic functions and OpenAI Gym tasks. Empirical results demonstrate that value suppression improves optimization efficiency and convergence under noise, making it a promising approach for noisy DFO problems. The chapter concludes with a discussion on the mechanism's potential applicability in noise-free environments.

In the previous chapters, we introduced the RACOS algorithm based on the sampling-and-learning (SAL) and sampling-and-classification (SAC) frameworks for derivative-free optimization (DFO). While RACOS has shown effectiveness in solving optimization problems in noise-free environments, many real-world applications involve noisy objective functions, where the evaluation of a solution is subject to random perturbations. In this chapter, we focus on extending RACOS to handle optimization problems under noise.

There are two popular mechanisms to handle noise in DFO: sampling and threshold selection equipped with re-evaluation. Sampling is a straightforward approach to reduce noise [1], where a given solution is independently evaluated multiple times, and the average of the noisy function values is used to approximate the true function value. However, obtaining an accurate estimate of the true function value requires a large sample size, which can be computationally expensive. Threshold selection equipped with re-evaluation [3, 6–9] independently re-evaluates solutions whenever a comparison occurs and replaces an old solution only when the value of a new solution is better by at least a fixed or dynamic threshold. This mechanism delays noise handling to the comparison step and has been shown to be helpful for DFO [10]. Compared to sampling, threshold selection with re-evaluation requires less computational cost and appears to be more efficient.

Y. Yu et al., *Derivative-Free Optimization*, Machine Learning: Foundations, Methodologies, and Applications, https://doi.org/10.1007/978-981-96-5929-6_8

In this chapter, we present a generic, simple, yet efficient noise handling mechanism called *value suppression* [11], which can be integrated into most DFO methods. Value suppression delays noise handling even further than threshold selection with re-evaluation. It does nothing about noise until the best-so-far solution has not been improved for a certain period, at which point it suppresses the value of the best-so-far solution and continues the optimization process. Value suppression not only helps DFO methods keep updating their best-so-far solutions but also records some suppressed solutions, from which the best solution with a more reliable value can be selected. Compared to sampling and threshold selection with re-evaluation, value suppression may require the least computational cost while still being effective.

We integrate the value suppression mechanism into SRacos in Chap. 6, resulting in the suppressed SRacos (SSRacos) for optimization under noise. To compare value suppression with other mechanisms on SRacos, we conduct experiments on two synthetic functions and reinforcement learning control tasks in OpenAI Gym. Experimental results demonstrate that value suppression can significantly improve the performance of SRacos under noise compared to other mechanisms.

8.1 Value Suppression

The idea of value suppression arises from the observation that if an algorithm has not updated its best-so-far solution for a period, the observed value of the best-so-far solution is likely to be much smaller than its true value due to noise. Thus, we suppress the value of the best-so-far solution when it remains the best for a long period, allowing the algorithm to resume its search and find better solutions.

The principle of value suppression is to adjust the overestimated value towards the true value. One way to implement this is to re-sample the value of the solution a sufficient number of times in the presence of unbiased random noise. In other situations, the suppression can be implemented by multiplying a discount factor, for example.

This simple mechanism can help the algorithm keep generating new solutions with better observed values. However, when the optimization finishes, the algorithm needs to choose the best solution among those it has generated. Since most solutions have only a noisy observed value, we only choose the best solution from the suppressed solutions, whose values are more reliable.

The value suppression mechanism can be easily applied to DFO algorithms that keep track of the best-so-far solution. The framework of DFO with value suppression is shown in Algorithm 8.1. First, the algorithm samples a set of solutions S and evaluates them (line 1). Let S^+ denote the best k solutions in S (line 2). In the following loop, the algorithm generates a new solution based on S^+, evaluates it, and uses it to update S^+ (lines 4–5). If S^+ does not update for a period, it suppresses the values of the samples in S^+ by re-sampling and saves these suppressed solutions (lines 6–8). Finally, the algorithm returns the best among all the suppressed solutions (line 10).

Under mild conditions, we can prove that Algorithm 8.1 is convergent, indicating that value suppression does not hurt optimization and is effective.

Theorem 8.1 *(Convergence) For a DFO algorithm $\mathcal{A}$ that generates any solution $\boldsymbol{x} \in \mathcal{X}$ with non-zero probability, assume that the noise follows the same i.i.d. and unbiased distribution for all solutions, and the value suppression assigns the true value to the solution. Then, the algorithm $\mathcal{A}$ with value suppression is convergent under noise, i.e., with probability 1, it will eventually output the optimal solution $\boldsymbol{x}^*$.*

Proof By assumption, once the true optimal solution $\boldsymbol{x}^*$ is generated during optimization under noise, $\boldsymbol{x}^*$ could be better than the best-so-far solution with a non-zero probability. Thus, by Algorithm 8.1, $\boldsymbol{x}^*$ could be absorbed into S^+ with probability 1 after a sufficient number of steps. Let S' denote the set of suppressed solutions, initialized as $S' = \emptyset$. For a fixed maximum allowed non-update iterations u, there exists a non-zero probability that S^+ will not be updated during u iterations, and thus $\boldsymbol{x}^*$ could be further absorbed into S' with probability 1 after a sufficient number of steps. By Algorithm 8.1, solutions cannot be removed from S' once absorbed. Note that the value suppression step for solutions in S' discloses the true function values as assumed. Since the algorithm will finally return the best solution in S', i.e., $\boldsymbol{x}^*$, value suppression is convergent. □

8.2 The SSRacos Algorithm

We integrate the value suppression framework with the SRacos in Algorithm 6.1. SRacos maintains two sets of solutions: a good solutions set (positive set) and a bad solutions set (negative set). A binary classifier is trained based on these two solution sets to learn the potential high-quality region in the solution space. The learned region contains one selected good solution from the positive set and excludes all the bad solutions from the negative set. Then, a new solution is uniformly sampled from this learned region with high probability or uniformly sampled from the entire solution space with the remaining probability. SRacos evaluates this new solution and updates both the positive and negative sets accordingly.

We now show how to integrate the value suppression mechanism into SRacos, resulting in the suppressed SRacos (SSRacos). Since we can observe the positive set B^+ in SRacos, if B^+ does not update for a period, we suppress the solutions in B^+. In the end, the best solution among all the suppressed ones is returned as output. The procedure of SSRacos is presented in Algorithm 8.2. The set B^S is used to collect the suppressed solutions and is initialized as an empty set (line 2). During the loop, after generating a new solution and updating B^+, B^-, and $(\tilde{x}, \tilde{y})$, SSRacos checks if the positive set B^+ has been updated (line 6). If it does not update for u iterations, SSRacos re-samples the solutions in B^+, suppresses their values, and saves them

Algorithm 8.1 Value Suppression Framework for Derivative-Free Optimization

Require:
  f^N: Noisy objective function
Ensure:
S = generate a set of solutions and evaluate them
S^+ = best k solutions in S
while termination condition is not met **do**
  $\boldsymbol{x}$ = generate a new solution based on S^+
  evaluate $\boldsymbol{x}$ and use it to update S^+
  if S^+ does not update for a period **then**
    suppress the function values of solutions in S^+
  end if
end while
return the best among all the suppressed solutions

Algorithm 8.2 Suppressed SRacos (SSRacos)

Require: (extra input compared to SRacos)
  $u \in \mathbb{N}^+$: Maximum allowed non-update iterations
  $v \in \mathbb{N}^+$: Re-sample size
  α: Balancing parameter
  `Re-sample`: Re-sample sub-procedure
Ensure:
1: Initialize SRacos
2: $B^S = \emptyset$
3: **for** $t = r + 1$ to $N - v$ **do**
4:   $(\boldsymbol{x}, y)$ = generate a new solution as in SRacos
5:   Use $(\boldsymbol{x}, y)$ to update B^+, B^-, $(\tilde{\boldsymbol{x}}, \tilde{y})$ in SRacos
6:   **if** B^+ does not update for u iterations **then**
7:     **for** $(\boldsymbol{x}_i, y_i)$ in B^+ **do**
8:       $\hat{y}_i =$ `Re-sample`$(\boldsymbol{x}_i, v)$
9:       $y_i = (1 - \alpha) y_i + \alpha \hat{y}_i$
10:       $B^S = B^S \cup \{(\boldsymbol{x}_i, \hat{y}_i)\}$
11:       $t = t + v$
12:     **end for**
13:   **end if**
14: **end for**
15: Re-sample $(\tilde{\boldsymbol{x}}, \tilde{y})$ and put it in B^S
16: **return** $\arg\min_{(\boldsymbol{x}, \hat{y}) \in B^S} \hat{y}$

with their mean values $\hat{y}$ in B^S (lines 7–12). The `Re-sample`$(\boldsymbol{x}, n)$ sub-procedure computes $f^N(\boldsymbol{x})$ for n times independently and returns the mean value $\hat{y}$. After that, it re-samples the best-so-far solution $(\tilde{\boldsymbol{x}}, \tilde{y})$ and saves it in B^S (line 15). Finally, the best solution in B^S is returned (line 16).

8.3 Empirical Study

In this section, we empirically demonstrate the effectiveness of the value suppression mechanism in reducing the negative effects of noise and saving computational cost. We compare value suppression with other noise handling mechanisms by integrating them into SRACOS. Specifically, SSRACOS with the number of solutions in the positive set $|B^+| > 1$ is abbreviated as VS. SRACOS with $|B^+| > 1$ is abbreviated as MPS. SRACOS with $|B^+| = 1$ is abbreviated as NO_MPS. SSRACOS with $|B^+| = 1$ is abbreviated as VS+NO_MPS. SRACOS equipped with sampling is abbreviated as SAMPLING, where a solution is evaluated n times and the average value is used to approximate its true function value [1]. SRACOS equipped with re-evaluation is abbreviated as REEVAL, where a solution is independently re-evaluated whenever its function value is required [4, 7, 8]. SRACOS equipped with threshold selection is abbreviated as TS, where a solution is considered better than another only when its function value is better by at least a threshold τ [2, 3, 9]. SRACOS equipped with the combination of re-evaluation and threshold selection is abbreviated as REEVAL+TS.

We conduct experiments on both synthetic functions and reinforcement learning control tasks in OpenAI Gym to investigate the ability of these mechanisms to handle noise. Additive Gaussian noise is used to create a noisy environment for synthetic functions. The OpenAI Gym environment is considered noisy because a policy may receive different total rewards under different initial states (more details can be found in the subsection on controlling tasks in OpenAI Gym). In addition to the noise from the original environment, we also add extra Gaussian noise to observe the performance of the mechanisms under different noise levels. Moreover, we analyze the sensitivity of the hyper-parameter u, the maximum allowed non-update iterations, on OpenAI Gym tasks.

8.3.1 Synthetic Functions

We choose the Ackley and Sphere functions to investigate the noise handling ability of each mechanism. The definition of these functions can be found in previous chapters.

We choose dimension sizes $D = 100$ and 1000 for both functions in the experiments. To create a noisy environment, we use additive Gaussian noise, i.e., the noisy function $f^N(\boldsymbol{x}) = f(\boldsymbol{x}) + \mathcal{N}(0, \sigma^2)$. For the Ackley function, the standard deviation $\sigma = 0.1$, and for the Sphere function, $\sigma = 1$.

The parameters of the noise handling mechanisms are set as follows. For threshold selection, we set the threshold value $\tau = \sigma$, because a solution that passes the threshold may be truly better with high probability. For MPS, the number of positive solutions is set to 5, which is a trade-off between computational cost and the chance of keeping good solutions. For sampling, the sample size is set to 10, balancing the accuracy of function evaluation and computational cost. For value suppression, we

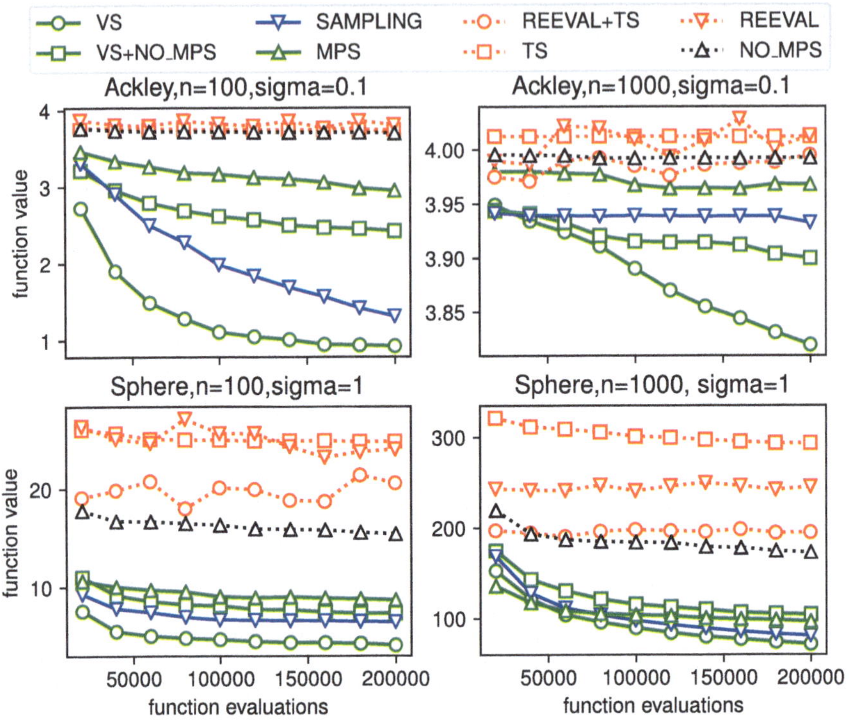

Fig. 8.1 Function value of each noise handling mechanism during the optimization process [11]. For the Ackley function, the dimension size $D = 100$ and 1000, and the standard deviation of noise $\sigma = 0.1$. For the Sphere function, $D = 100$ and 1000, and $\sigma = 1$

set the maximum allowed non-update iterations $u = 500$, the re-sample size $v = 100$, and the balancing parameter $\alpha = 0.5$. Other parameters are set to their default values.

For each setting with different dimension size D or standard deviation of noise σ, we run each mechanism 10 times independently to minimize the noisy function $f^N(\boldsymbol{x})$. The total number of function evaluations is set to 200,000. The true function value $f(\boldsymbol{x})$ of the best-found solution during the process is shown in Fig. 8.1, and the true function value of the returned best solution is listed in Table 8.1 SRACOS with the number of solutions in the positive set $|B^+| = 1$ (NO_MPS) is chosen as the baseline. The results show that VS achieves the best performance in all settings. SAMPLING and MPS show similar performance and are able to reduce the effects of noise. VS+NO_MPS performs better than MPS and is competitive with SAMPLING. However, REEVAL, TS, and REEVAL+TS are worse than the baseline or not significantly different. Figure 8.1 shows that VS requires significantly fewer iterations to achieve good performance compared to the other mechanisms. Specifically, on the Ackley function with dimension size 100 and noise level 0.1, VS only needs less than half the function evaluations to reach a function value below 2 compared

Table 8.1 Function value for each noise handling mechanism [11]. For the Ackley function, the dimension size $D = 100$ and 1000, and the standard deviation of noise $\sigma = 0.1$. For the Sphere function, $D = 100$ and 1000, and $\sigma = 1$. The number of function evaluations is set to 200,000

Function_DimSize_Noise	VS	VS+NO_MPS	SAMPLING	MPS	REEVAL+TS	REEVAL	TS	NO_MPS
Ackley_100_0.1	**0.93**	2.43	1.32	2.95	3.71	3.82	3.75	3.69
Ackley_1000_0.1	**3.82**	3.90	3.93	3.96	3.99	4.01	4.01	3.99
Sphere_100_1	**4.17**	7.41	6.53	8.74	20.65	24.07	24.88	15.41
Sphere_1000_1	**72.41**	104.81	81.78	97.16	196.14	246.22	294.42	172.98

to the second best mechanism, SAMPLING. This indicates that the presented VS mechanism can significantly reduce the computational and time cost.

8.3.2 *Controlling Tasks in OpenAI Gym*

OpenAI Gym provides a toolkit for reinforcement learning research.[1] We choose the following controlling tasks to compare the noise handling ability of each mechanism: Acrobot, MountainCar, HalfCheetah, Humanoid, Swimmer, Ant, Hopper, and LunarLander.

We use the framework of direct policy search to solve these tasks. Direct policy search applies optimization algorithms to search the parameter space of a policy, which is often represented by a neural network [5]. The objective is to maximize the accumulated reward of a policy. Specifically, a policy is represented by a neural network with an input layer for the observation of the state, an output layer for the available actions, and several hidden layers. In each step, an agent receives an observation of the state and takes an action according to its policy. It then receives the reward for that action together with the observation of the next state. This interaction is repeated until the maximum number of steps is reached or the game is over. The accumulated reward is used to evaluate the performance of a policy. The agent may receive different accumulated rewards if the initial state is different. Therefore, we consider the environment to be noisy. To summarize, our goal is to find the best parameters $\boldsymbol{w}$ for the neural network to achieve the best performance. The difficulty lies in the fact that the accumulated reward $f^N(\boldsymbol{w})$ used to evaluate the performance can be noisy during optimization. Thus, we use the noise handling mechanisms to reduce the effect of noise in this environment and compare their performances. The settings of the neural network and OpenAI Gym tasks are listed in Table 8.2, where d_{state}, #Actions, NN nodes, #Weights, and Horizon denote the dimension of the observation, the dimension of the action, the hidden layers of the neural network, the total number of parameters in the neural network, and the maximum number of steps, respectively.

We compare these mechanisms under the same parameter settings of SRacos, which are listed in Table 8.3, where $\#B^-$ and $\#B^+$ denote the size of the negative set and positive set, respectively, and U-bits denotes the number of bits that can be changed when generating a new solution from a positive solution. From the experimental results on synthetic functions, we note that VS achieves the best performance. Thus, we combine the other mechanisms with MPS to see if they can improve the performance of MPS better than value suppression. On OpenAI Gym tasks, the total number of function evaluations is set to 20,000.

The parameters of these mechanisms are set as follows. For sampling, the sample size is set to 10. For threshold selection, the noise level is estimated to choose a proper threshold value. To estimate the standard deviation of the noise, we draw 10

[1] https://gym.openai.com.

Table 8.2 Parameters of the OpenAI Gym tasks [11]

Task	d_{state}	#Actions	NN nodes	#Weights	Horizon
Acrobot-v1	6	1	5, 3	48	500
MountainCar-v0	2	1	5	15	200
HalfCheetah-v1	17	6	10	230	1,000
Humanoid-v1	376	17	25	9,825	1,000
Swimmer-v1	8	2	5, 3	61	1,000
Ant-v1	111	8	15	1,785	1,000
Hopper-v1	11	3	9, 5	159	1,000
LunarLander-v2	8	1	5, 3	58	1,000

Table 8.3 Parameters of SRACOS and noise level [11]

Task	$\#B^-$	$\#B^+$	U-bits	Noise level
Acrobot-v1	20	2	1	28.0
MountainCar-v0	20	2	1	10.0
HalfCheetah-v1	50	3	3	200.0
Humanoid-v1	20	2	3	56.0
Swimmer-v1	50	4	2	10.0
Ant-v1	20	2	3	46.0
Hopper-v1	50	6	4	60.0
LunarLander-v2	50	5	3	50.0

samples from the solution space, evaluate each sample 1,000 times independently, and compute the standard deviation. The average standard deviation of these 10 samples is used to estimate the standard deviation of the noise. The values are listed in Table 8.3 as the noise level σ. We round these estimated values to the nearest integers and set the threshold value $\tau = \sigma$. For value suppression, we set the maximum allowed non-update iterations $u = 500$, the re-sample size $v = 100$, and the balancing parameter $\alpha = 0.5$.

For each mechanism, the optimization algorithm is run independently 10 times. At the end of each run, the average accumulated reward of 1,000 simulations is used to estimate the performance of the found policy. The mean and standard deviation of the 10 policies are reported in Table 8.4, where the standard deviation of additional Gaussian noise is set to 0, 0.1, and 1 times the noise level in Table 8.3, respectively. The mean value of a mechanism is in bold if it is not significantly worse than the mechanism with the maximal mean value under a t-test. SRACOS with the number of solutions in the positive set $|B^+| > 1$ (MPS) is chosen as the baseline for comparison. We can observe that VS performs the best on all tasks, while SAMPLING and REEVAL+TS achieve the best performance only on the Humanoid-v1 task. REEVAL, TS, and REEVAL+TS perform worse than the baseline on some tasks. Since VS achieves the best performance within a given solution evaluation budget,

Table 8.4 Mean value and standard deviation of reward achieved by each mechanism [11]. The mark ↓ means that a smaller reward is better for the task, while ↑ means that a larger reward is better. The mean value of a mechanism is in bold if it is not significantly worse than the mechanism with the maximal mean value under a t-test with a significance level of $\gamma = 10\%$

Noise	Task	VS	SAMPLING	REEVAL+TS	REEVAL	TS	MPS
0	Acrobot-v1↓	**80.76**±1.38	85.51±3.46	94.03±13.10	140.58±120.76	121.74±40.45	86.50±4.45
	MountainCar-v0↓	**134.92**±3.87	141.94±8.29	181.07±22.61	199.86±0.40	158.67±17.10	150.85±13.33
	HalfCheetah-v1↑	**1924.60**±278.08	1408.54±383.85	1231.46±209.50	752.38±346.83	968.50±427.66	1388.27±479.94
	Humanoid-v1↑	**461.85**±23.92	**459.53**±22.81	**473.05**±34.70	444.60±39.12	433.87±32.57	422.40±41.84
	Swimmer-v1↑	**360.51**±3.45	342.02±20.25	355.40±3.38	348.84±9.70	336.94±16.33	289.97±71.70
	Ant-v1↑	**1312.85**±90.16	1130.54±55.35	1052.64±95.64	1056.09±78.96	1016.05±36.28	1126.89±123.11
	Hopper-v1↑	**1111.91**±117.69	1002.03±48.55	1003.73±12.84	641.23±406.58	630.02±242.86	873.87±186.46
	LunarLander-v2↑	**80.40**±54.51	−21.23±91.65	−191.23±45.77	−185.59±24.45	−172.36±97.17	−187.70±107.00
0.1	Acrobot-v1↓	**81.26**±1.44	**84.45**±5.48	92.91±11.18	117.18±61.64	111.94±51.53	88.87±5.12
	MountainCar-v0↓	**140.18**±9.20	**154.56**±24.52	172.97±22.88	200.00±0.04	163.45±21.21	158.65±16.69
	HalfCheetah-v1↑	**1603.95**±469.26	**1314.68**±674.56	1025.08±372.57	572.12±755.55	573.43±687.99	1228.45±579.65
	Humanoid-v1↑	**460.15**±25.12	426.24±19.89	**464.92**±28.66	418.89±45.39	396.49±39.76	426.85±21.51
	Swimmer-v1↑	**361.42**±2.38	356.64±4.22	341.74±18.11	332.89±40.37	294.99±54.03	323.52±39.40
	Ant-v1↑	**1179.63**±93.48	1097.92±78.16	1006.45±22.73	1007.23±18.04	998.59±3.10	1097.26±95.47
	Hopper-v1↑	**1098.84**±92.24	1015.18±38.21	998.98±13.98	797.96±288.05	512.60±273.42	545.55±251.03
	LunarLander-v2↑	**40.39**±65.07	−38.60±69.64	−180.50±25.73	−220.36±81.02	−250.69±107.65	−240.22±79.50
1	Acrobot-v1↓	**84.28**±2.05	86.89±2.54	96.53±12.12	162.16±119.12	126.18±52.74	123.88±77.87
	MountainCar-v0↓	**136.99**±5.02	167.77±27.89	162.97±22.92	200.02±0.34	175.12±24.49	175.80±16.00
	HalfCheetah-v1↑	**1286.29**±440.46	858.87±468.68	776.81±400.41	377.91±585.26	585.48±611.87	**987.11**±424.63
	Humanoid-v1↑	**470.63**±36.79	**433.63**±39.01	**452.51**±37.53	395.58±57.60	406.76±43.01	425.62±33.90
	Swimmer-v1↑	**355.73**±5.86	336.34±38.57	340.85±16.27	334.31±12.90	302.17±35.86	335.54±20.73
	Ant-v1↑	**1010.00**±15.27	997.07±3.57	**1000.48**±9.38	992.90±3.55	993.30±2.98	994.67±2.64
	Hopper-v1↑	**996.14**±70.48	**1010.32**±30.11	767.96±350.41	314.72±339.94	692.43±190.35	769.20±285.23
	LunarLander-v2↑	**58.02**±74.66	−81.39±73.79	−175.45±18.49	−196.36±70.33	−264.02±126.33	−258.27±91.00

it requires the least computational and time cost compared to the other mechanisms, making it the most efficient.

We further add Gaussian noise to these tasks and observe the performance of the mechanisms under additional noise. Two experiments are conducted with different levels of extra noise. For the first one, the standard deviation of the additional Gaussian noise is set to 0.1 times the noise level in Table 8.3. For the second one, the standard deviation is set to 1 times the noise level. We keep the other parameters the same as in the previous experiment for OpenAI Gym. SRACOS with the number of solutions in the positive set $|B^+| > 1$ (MPS) is chosen as the baseline for comparison. The results are listed in Table 8.4, and the comparison of the mechanisms under different extra noise levels is shown in Fig. 8.2. We can observe that VS achieves the best or equally best performance under the t-test in all tasks. Although it performs worse than the environment without extra noise in tasks like Ant-v1 and HalfCheetah-v1, it achieves almost the same or even better performance in other tasks. However, SAMPLING and REEVAL+TS perform worse as the extra noise increases in most tasks. In tasks like HalfCheetah-v1 and Swimmer-v1, they do not perform better than the baseline, which does not handle noise.

8.3.3 *Hyper-Parameter Analysis*

We also investigate the sensitivity of the hyper-parameter u, i.e., the maximum allowed non-update iterations, in OpenAI Gym. The experimental setting is the same as that without extra noise, and u is chosen from $\{100, 500, 1000\}$. In the previous experiments, u was always set to 500. The results are shown in Table 8.5.

Table 8.5 indicates that the results are not significantly different in most tasks when $u \in \{100, 500, 1000\}$. This implies that the hyper-parameter u is not very sensitive. If u is too small, the confidence that the solution is trapped due to noise is low, and it may waste samples to accurately evaluate a solution that would be replaced soon. If u is too large, the confidence is high, but it may waste samples waiting for the confidence to build up. Therefore, the choice of u should be balanced. According to the experimental results, the default setting $u = 500$ should be suitable in many cases.

8.4 Summary

In many real-world applications, such as policy search in reinforcement learning, the environment is noisy, and noise can significantly degrade the performance of derivative-free optimization methods. This chapter presents a generic, simple, yet effective noise handling mechanism called value suppression [11]. Value suppression can be integrated into most derivative-free optimization methods to handle and reduce noise. To verify the effectiveness of this mechanism, we integrate it into

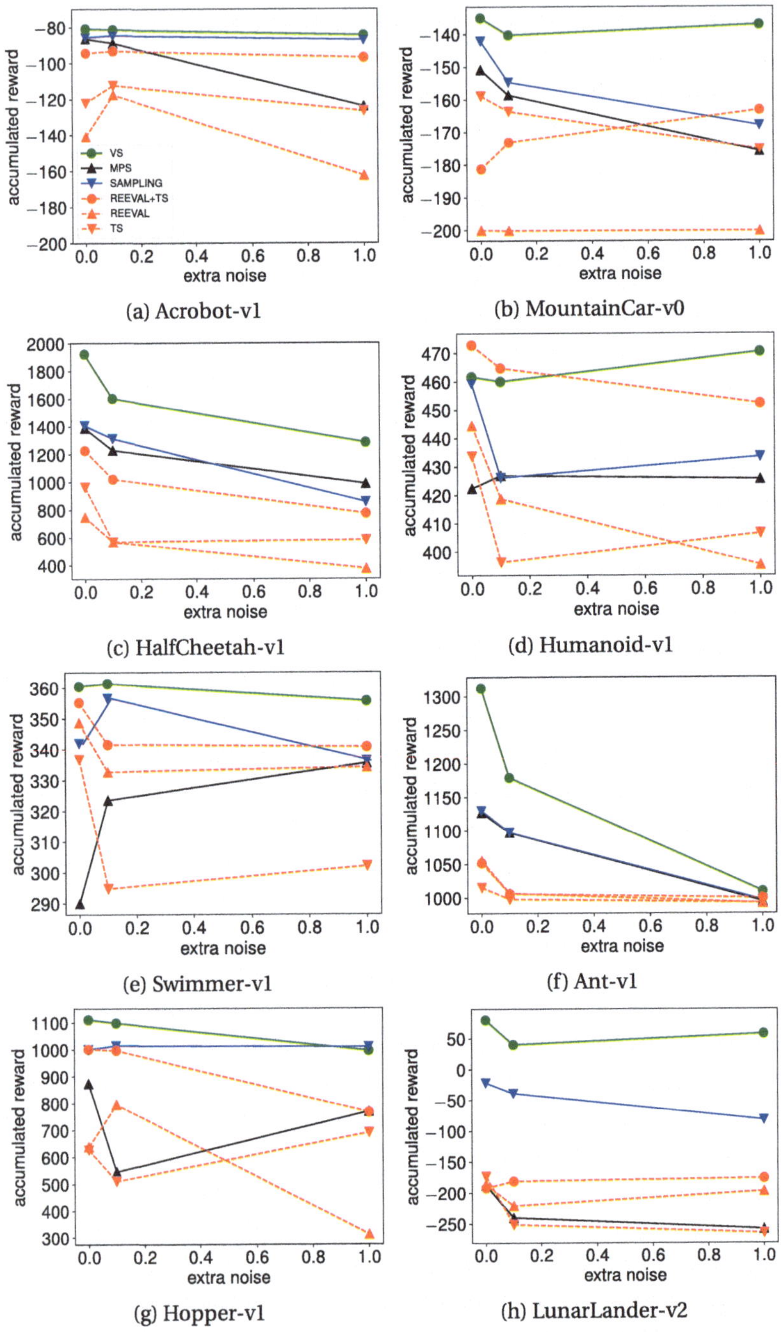

(a) Acrobot-v1

(b) MountainCar-v0

(c) HalfCheetah-v1

(d) Humanoid-v1

(e) Swimmer-v1

(f) Ant-v1

(g) Hopper-v1

(h) LunarLander-v2

Fig. 8.2 Comparison of the performance under extra noise levels of 0, 0.1, and 1 times the noise level, respectively [11]

Table 8.5 Hyper-parameter analysis of maximum allowed non-update iterations u [11], where $u \in \{100, 500, 1000\}$. The mark ↓ means that a smaller reward is better for the task, while ↑ means that a larger reward is better. The mean value of a mechanism is in bold if it is not significantly worse than the mechanism with the maximal mean value under a t-test with a significance level of $\gamma = 10\%$

Task	500	100	1000
Acrobot-v1↓	**80.76**±1.38	**79.52**±2.54	82.06±1.39
MountainCar-v0↓	**134.92**±3.87	**132.30**±4.28	**134.96**±4.52
HalfCheetah-v1↑	**1924.60**±278.08	1554.27±486.50	**1773.89**±548.06
Humanoid-v1↑	**461.85**±23.92	**460.39**±34.97	**455.46**±35.26
Swimmer-v1↑	**360.51**±3.45	**360.91**±2.33	**359.35**±5.09
Ant-v1↑	**1312.85**±90.16	**1239.24**±119.53	1181.04±91.45
Hopper-v1↑	**1111.91**±117.69	**1046.35**±27.49	**1058.87**±30.77
LunarLander-v2↑	**80.40**±54.51	−23.02±62.22	**21.04**±80.90

SRACOS, resulting in the suppressed SRACOS (SSRACOS). Experimental results on both synthetic functions and reinforcement learning control tasks in OpenAI Gym demonstrate that value suppression can perform better than other popular noise handling mechanisms, such as sampling and threshold selection with re-evaluation. In the future, we will further explore whether value suppression can be helpful in noise-free environments. Intuitively, value suppression may help the algorithm escape local optima even in the absence of noise.

References

1. Arnold DV, Beyer H (2006) A general noise model and its effects on evolution strategy performance. IEEE Trans Evol Comput 10(4):380–391
2. Bartz-Beielstein T (2005) Evolution strategies and threshold selection. In: Proceedings of the 2nd international workshop on hybrid metaheuristics, Barcelona, Spain, pp 104–115
3. Beielstein T, Markon S (2002) Threshold selection, hypothesis tests, and DOE methods. In: Proceedings of the 2002 IEEE congress on evolutionary computation, Honolulu, HI, pp 777–782
4. Doerr B, Hota A, Kötzing T (2012) Ants easily solve stochastic shortest path problems. In: Proceedings of the 14th ACM annual conference on genetic and evolutionary computation, Philadelphia, PA, pp 17–24
5. El-Fakdi A, Carreras M, Ridao P (2006) Towards direct policy search reinforcement learning for robot control. In: Proceedings of the 2006 IEEE/RSJ international conference on intelligent robots and systems Beijing, China, pp 3178–3183
6. Gießen C, Kötzing T (2016) Robustness of populations in stochastic environments. Algorithmica 75(3):462–489
7. Goh CK, Tan KC (2007) An investigation on noisy environments in evolutionary multiobjective optimization. IEEE Trans Evol Comput 11(3):354–381
8. Jin Y, Branke J (2005) Evolutionary optimization in uncertain environments-A survey. IEEE Trans Evol Comput 9(3):303–317

9. Markon S, Arnold DV, Bäck T, Beielstein T, Beyer H (2001) Thresholding-A selection operator for noisy ES. In: Proceedings of the 2001 IEEE Congress on Evolutionary Computation, Seoul, South Korea, pp 465–472
10. Qian C, Yu Y, Zhou Z (2018) Analyzing evolutionary optimization in noisy environments. Evol Comput 26(1):1–41
11. Wang H, Qian H, Yu Y (2018) Noisy derivative-free optimization with value suppression. In: Proceedings of the 32nd AAAI conference on artificial intelligence, New Orleans, LA

Chapter 9
Optimization with Parallel Computing

Abstract This chapter introduces ASRacos, an asynchronous variant of the SRacos algorithm, designed to accelerate derivative-free optimization through parallel computing. While SRacos excels in sequential optimization, its structure limits parallelization, which is crucial for time-consuming tasks. ASRacos modifies SRacos to enable asynchronous parallelism, allowing multiple servers to evaluate solutions concurrently while maintaining the sequential update structure. The chapter provides a theoretical analysis of ASRacos, including its query complexity and conditions under which it outperforms SRacos. Empirical studies compare ASRacos with other parallel methods on synthetic functions and reinforcement learning tasks, demonstrating its ability to achieve near-linear speedup and superior solution quality. The results highlight the effectiveness of asynchronous parallelism in accelerating optimization without compromising performance. Future work may explore integrating noise-handling methods and applying ASRacos to large-scale real-world problems.

While SRACOS has shown outstanding performance in various applications [7–9], its sequential structure prevents it from being parallelized, which can be a limitation for time-consuming optimization tasks. Asynchronous parallelism is an effective way to accelerate optimization, but it can destroy the sequential structure of optimization algorithms, potentially deteriorating their performance. However, some optimization algorithms have been proven to preserve their performance under asynchronous parallelization, such as stochastic gradient descent for first-order optimization of differentiable functions [10] and Pareto optimization for zeroth-order optimization in binary space [5].

In this chapter, we present an asynchronous variant of SRACOS called ASRACOS [4]. We apply a feasible modification to SRACOS to make it parallelizable and implement its asynchronous version ASRACOS, which maintains the sequential structure while being able to utilize multiple servers. We provide the (ϵ, δ)-query complexity bound of ASRACOS in theoretical analysis and further give the condition under which ASRACOS can achieve better (or worse) performance than SRACOS, even when using the same number of evaluations. We empirically compare ASRACOS with several other parallel classification-based optimization algorithms on four synthetic testing

Y. Yu et al., *Derivative-Free Optimization*, Machine Learning: Foundations, Methodologies, and Applications, https://doi.org/10.1007/978-981-96-5929-6_9

functions and apply them to direct policy search for six reinforcement learning control tasks, where an artificial neural network is used as the policy and optimized. Experimental results show that ASRACOS can achieve almost linear speedup while preserving good solution quality.

9.1 The Asynchronous SRACOS (ASRACOS) Algorithm

The idea of making SRACOS parallelizable is straightforward: sample N_s solutions (where N_s is the number of evaluation servers) after initialization, rather than a single solution. These solutions can then be evaluated in parallel. Whenever an evaluation is finished, the method updates the model and samples the next solution for evaluation. Note that the sequential update structure is still maintained through this modification.

The pseudocode of ASRACOS is shown in Algorithm 9.1. The notations and subprocedures used in the algorithm are the same as those introduced in Chap. 6 for the SRACOS algorithm.

Algorithm 9.1 Asynchronous SRACOS (ASRACOS)

Require: (extra input compared to SRACOS)
  $N_s \in \mathbb{N}^+$: The number of evaluation servers
Ensure:
Collect $S = \{\boldsymbol{x}_1, \ldots, \boldsymbol{x}_r\}$ by i.i.d. sampling from $\mathcal{U}_\mathcal{X}$
$B = \{(\boldsymbol{x}_1, y_1), \ldots, (\boldsymbol{x}_r, y_r)\}, \forall \boldsymbol{x}_i \in S : y_i = f(\boldsymbol{x}_i)$
$(B^+, B^-) = \texttt{Selection}(B; k)$
$h_1 = \mathcal{C}(B^+, B^-)$
$D, E = \mathit{SharedQueue}\{\ \}, \mathit{SharedQueue}\{\ \}$
$D = \{\boldsymbol{x}_{r+1}, \ldots, \boldsymbol{x}_{r+N_s}\} = \lambda\text{-}\texttt{Sampling}_{N_s}(\mathcal{U}_{D_{h_1}}, \mathcal{U}_\mathcal{X})$
Run `Evaluation(D, E)` sub-procedures on N_s daemon threads
for $t = r + 1$ to N **do**
  $(\boldsymbol{x}, y) = \texttt{take}(E)$
  $[(\boldsymbol{x}', y'), B^+] = \texttt{Replace}((\boldsymbol{x}, y), B^+, \text{'strategy\_P'})$
  $[\emptyset, B^-] = \texttt{Replace}((\boldsymbol{x}', y'), B^-, \text{'strategy\_N'})$
  $(\tilde{\boldsymbol{x}}, \tilde{y}) = \arg\min_{(\boldsymbol{x}, y) \in B^+ \cup \{(\tilde{\boldsymbol{x}}, \tilde{y})\}} y$
  $h = \mathcal{C}(B^+, B^-)$
  $\boldsymbol{x} = \lambda\text{-}\texttt{Sampling}_1(\mathcal{U}_{D_h}, \mathcal{U}_\mathcal{X})$
  $\texttt{put}(\boldsymbol{x}, D)$
end for
return $(\tilde{\boldsymbol{x}}, \tilde{y})$

`Evaluation(D, E)`:
while true **do**
  $\boldsymbol{x} = \texttt{take}(D)$
  $y = f(\boldsymbol{x})$
  $\texttt{put}((\boldsymbol{x}, y), E)$
end while

After initialization, ASRACOS obtains two tuple sets B^+ and B^- according to the function values (line 3). Then, a binary classifier is trained based on these two sets to learn the potential high-quality region in the solution space (line 4). The learned region contains one selected good solution from the positive set and excludes all the bad solutions from the negative set. ASRACOS contains two first-in-first-out blocking queues: D for the unevaluated solutions and E for the evaluated solutions. D and E are shared between the main thread and the evaluation threads for data communication. D is initialized with the first batch of sampled solutions, and E is initialized as empty (lines 5–6). Then, ASRACOS starts N_s evaluation servers (implemented as newly created threads), each continuously evaluating a solution taken from D and putting the result $(\boldsymbol{x}, y)$ into E (lines 21–23). In the following loop, ASRACOS takes the evaluated tuple $(\boldsymbol{x}, y)$ from E and uses it to update the tuple sets B^+ and B^- (lines 9–11). Once a new binary classifier $\mathcal{C}$ is trained (line 13), a new solution is sampled and put into D (lines 14–15).

In summary, ASRACOS divides the sequential evaluation and update procedure of SRACOS into two components: the asynchronous evaluation component and the sequential model update component. The asynchronous evaluation component can utilize multiple servers, while the model update component can still update the classification model sequentially, maintaining the sequential structure of SRACOS. The blocking queues D and E are created for data communication between threads.

Figure 9.1 demonstrates the flowcharts of the optimization procedures of ASRACOS and SRACOS, where the solid arrow denotes the sampling and evaluation procedure, the hollow arrow denotes an update on the data distribution D_t, and s_i, s_j, and s_k denote the unused solutions sampled previously. It can be observed that D_t is always updated by the solution sampled from D_t for SRACOS, while it can be updated by the solution sampled from another distribution several iterations ago for ASRACOS, which causes the difference in the data distributions of the two algorithms. The next section discusses the effect of this difference on the query complexity of ASRACOS.

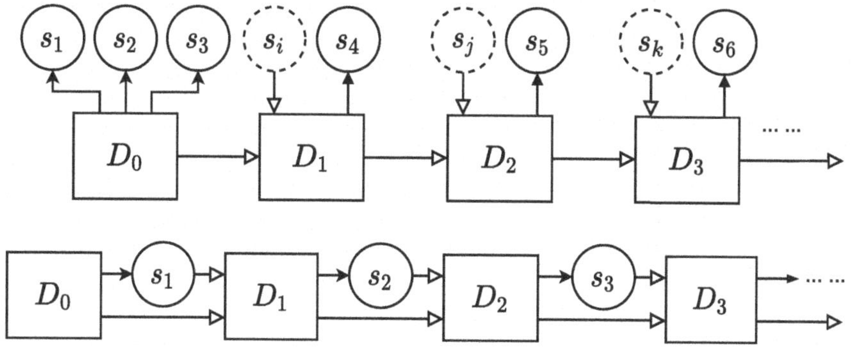

Fig. 9.1 Flowcharts of the optimization procedures of ASRACOS (top, using three evaluation servers) and SRACOS (bottom) [4]

9.2 Theoretical Analysis

We derive an upper bound on the (ϵ, δ)-query complexity of ASRACOS under the conditions of error-target θ-dependence and γ-shrinking rate, as introduced in Chap. 4.

Lemma 9.1 *Given an objective function f, $\epsilon > 0$, and $0 < \delta < 1$, if* ASRACOS *has error-target θ-dependence and γ-shrinking rate, then its (ϵ, δ)-query complexity is upper bounded by*

$$O\left(\max\left\{\frac{1}{\mu(D_\epsilon)}\left((1-\lambda)+\frac{\lambda}{\gamma(N-r)}\sum_{t=r+1}^{N}\Phi_t^A\right)^{-1}\ln\frac{1}{\delta}, N\right\}\right), \tag{9.1}$$

where $\Phi_t^A = \left(1 - \epsilon_{D_t^A} - \sqrt{2D_{KL}(D_t^A\|\mathcal{U}_\mathcal{X})} - \theta\right)\cdot \mu(D_{\alpha_t})^{-1}$, *and* $|\mathcal{X}|$ *is the volume of* $\mathcal{X}$.

The proof of Lemma 9.1 is similar to the proof of Theorem 4.3 in Chap. 4, except for the values of ϵ_{D_t} and $D_{KL}(D_t\|\mathcal{U}_\mathcal{X})$ at each iteration. Using Lemma 9.1, we can compare the query complexity bounds of ASRACOS and SRACOS. The result is shown in Theorem 9.1.

Theorem 9.1 *Ignoring the constant factor and fixing θ and γ,* ASRACOS *can have a better (or worse) query complexity upper bound than* SRACOS *if, for any iteration t,*

$$\epsilon_{D_t^A} - \epsilon_{D_t^S} < (>) \sqrt{2D_{KL}(D_t^S\|\mathcal{U}_\mathcal{X})} - \sqrt{2D_{KL}(D_t^A\|\mathcal{U}_\mathcal{X})}. \tag{9.2}$$

Proof Let D_t^A and D_t^S denote the distributions under which the classifiers are trained in iteration t of ASRACOS and SRACOS, respectively, and $\epsilon_{D_t^A}$ and $\epsilon_{D_t^S}$ denote their corresponding generalization errors.

Recall the (ϵ, δ)-query complexity bound of a classification-based sequential DFO algorithm derived in Theorem 4.3 of Chap. 4. Given an objective function f, $\epsilon > 0$, and $0 < \delta < 1$, if a classification-based sequential DFO algorithm has error-target θ-dependence and γ-shrinking rate, then its (ϵ, δ)-query complexity is upper bounded by

$$O\left(\max\left\{\frac{1}{\mu(D_\epsilon)}\left((1-\lambda)+\frac{\lambda}{\gamma(N-r)}\sum_{t=r+1}^{N}\Phi_t^S\right)^{-1}\ln\frac{1}{\delta}, N\right\}\right), \tag{9.3}$$

where $\Phi_t^S = \left(1 - \epsilon_{D_t^S} - \sqrt{2D_{KL}(D_t^S\|\mathcal{U}_\mathcal{X})} - \theta\right)\cdot \mu(D_{\alpha_t})^{-1}$, and $|\mathcal{X}|$ is the volume of $\mathcal{X}$.

In Lemma 9.1, ignoring the constant factor and letting $\epsilon > 0$ be small enough, we only need to focus on the term

$$\frac{1}{\mu(D_\epsilon)}\left((1-\lambda)+\frac{\lambda}{\gamma(N-r)}\sum_{t=r+1}^{N}\Phi_t^A\right)^{-1}\ln\frac{1}{\delta}, \tag{9.4}$$

where $\Phi_t^A = \left(1-\epsilon_{D_t^A}-\sqrt{2D_{KL}(D_t^A\|\mathcal{U}_\mathcal{X})}-\theta\right)\cdot\mu(D_{\alpha_t})^{-1}$.

Comparing Lemma 9.1 and Eq. (9.3), to compare ASRACOS with SRACOS, it suffices to compare the terms $1-\epsilon_{D_t^A}-\sqrt{2D_{KL}(D_t^A\|\mathcal{U}_\mathcal{X})}-\theta$ and $1-\epsilon_{D_t^S}-\sqrt{2D_{KL}(D_t^S\|\mathcal{U}_\mathcal{X})}-\theta$, ignoring the corresponding constant factors. It can be verified that, for any iteration t, if $\epsilon_{D_t^A}-\epsilon_{D_t^S}<\sqrt{2D_{KL}(D_t^S\|\mathcal{U}_\mathcal{X})}-\sqrt{2D_{KL}(D_t^A\|\mathcal{U}_\mathcal{X})}$, then ASRACOS has a better query complexity upper bound than SRACOS; if $\epsilon_{D_t^A}-\epsilon_{D_t^S}>\sqrt{2D_{KL}(D_t^S\|\mathcal{U}\mathcal{X})}-\sqrt{2DKL(D_t^A|\mathcal{U}_\mathcal{X})}$, then ASRACOS is worse. □

Theorem 9.1 reveals that if the difference in the training distributions between the two algorithms has a greater influence than the difference in generalization error, ASRACOS can be better than SRACOS even when using the same number of evaluations. Moreover, ASRACOS can use nearly N_s times more evaluations than SRACOS within the same time. Therefore, it is much easier for ASRACOS to find a better solution than SRACOS in practice.

9.3 Empirical Study

We evaluate the performance of ASRACOS in two environments. One is the optimization of classical synthetic functions, containing a convex function and three highly non-convex functions; the other is the controlling tasks in OpenAI Gym, an open source environment for reinforcement learning research.

We investigate the performance of the asynchronous parallelism on the classification-based optimization algorithms, including convergence rate, speedup ratio, and solution quality. We compare our method with another two parallel classification-based methods: Parallel RA-COS (PRACOS) and Parallel SRACOS (PSRACOS). PRACOS is a simple parallel implementation of the batch-mode method SRACOS [6]. PSRACOS shares the same structure with ASRACOS, and only varies in that the classification model will not update until the slowest evaluation server finishes evaluation. Note that when the number of evaluation servers is 1, ASRACOS and PSRACOS are equivalent to SRACOS, and PRACOS equals SRACOS. Reference [3] has compared the performance of a sequential classification-based optimization algorithm with other state-of-the-art derivative-free optimization algorithms, so we omit these comparisons in this chapter.

9.3.1 On Synthetic Functions

We choose four benchmark testing functions: the convex Sphere function and the highly non-convex Ackley, Rastrigin, and Griewank function. They are defined as

$$Sphere(\boldsymbol{x}) = \sum_{i=1}^{d} x_i^2 , \tag{9.5}$$

$$Ackley(\boldsymbol{x}) = -20e^{-\frac{1}{5}\sqrt{\frac{1}{d}\sum_{i=1}^{d} x_i^2}} - e^{\frac{1}{d}\sum_{i=1}^{d}\cos(2\pi x_i)} + 20 + e , \tag{9.6}$$

$$Rastrigin(\boldsymbol{x}) = 10d + \sum_{i=1}^{d} \left[{x_i}^2 - 10 \cos(2\pi x_i) \right] , \tag{9.7}$$

$$Griewank(\boldsymbol{x}) = \sum_{i=1}^{d} \frac{x_i^2}{4000} - \prod_{i=1}^{d} \cos(\frac{x_i}{\sqrt{i}}) + 1 . \tag{9.8}$$

The functions are minimized within the solution space $X = [-1, 1]^d$, of which the minimum value is 0 and the optimal solution is $(0, \ldots, 0)$. In the implementation, we choose d to be 100 and shift the optimal solution by 0.2, which means the new optimal solution is $(0.2, 0.2, \ldots, 0.2)$, to avoid possible optimization bias to the origin point. In addition, we add a fixed 1-s sleep for each evaluation. This is a reasonable modification since any distributed algorithm faces the networking overhead. If the evaluation time cost is even smaller than the networking overhead, parallelization may not be necessary. Another 1-s sleep with 0.25 probability is also added to simulate a situation where evaluation servers vary in computational performance, i.e., some servers are explicitly slower than others, which is common in real-world applications. Each algorithm is repeated 10 times independently, and the average performance is reported.

On Convergence Rate
We firstly study the convergence rate of ASRACOS. We set the time for optimization to be 2000 s compare the performance with the number of evaluation servers $N_s = 1, 2, 4, 6, 8, 10$. The results are shown in Fig. 9.2. The dotted line represents the optimal value that ASRACOS obtains when using one server (also the result of SRACOS). It can be observed that ASRACOS with more evaluation servers reduces the objective function value with a higher rate, indicating that asynchronous parallelism can accelerate the convergence.

On Speedup
We then study the speedup w.r.t the number of evaluation servers (N_s). We set the budget to be 2000 for each algorithm and calculate the speedup as $S_i = T_1 / T_i$, where T_i represents the time consumed when $N_s = i$. The results are shown in Fig. 9.3. From the left plots of each function, we can observe that ASRACOS (blue line) achieves linear speedup, notably better than PRACOS and PSRACOS. The results reflect the

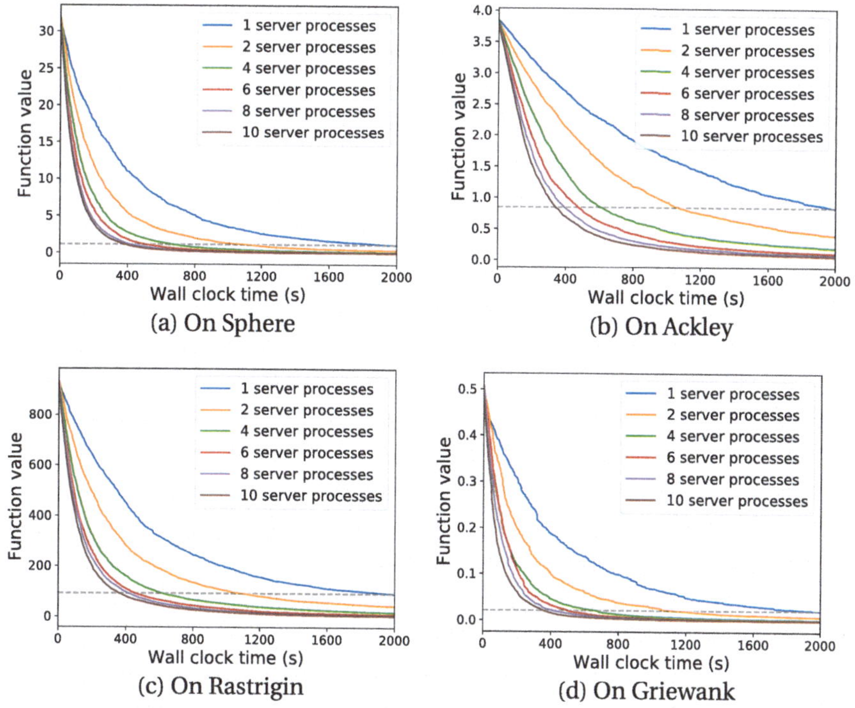

Fig. 9.2 Comparison of the convergence rate with the number of evaluation servers $N_s = 1, 2, 4, 6, 8, 10$ [4]

advantage of asynchronous parallelism over simple parallelism when servers vary in computational performance.

On Solution Quality

To study the solution quality w.r.t. the number of evaluation servers within the same time constraint, we set the time for optimization to be 20 min for each algorithm. The results are shown in the right plots of Fig. 9.3. We can see that algorithms using more servers get better solution quality and ASRacos achieves the best performance among them.

9.3.2 On Controlling Tasks in OpenAI Gym

OpenAI Gym is a toolkit for developing and comparing reinforcement learning algorithms. The toolkit provides many controlling tasts, from which we choose Acrobot, MountainCar, Pendulum, HalfCheetah, Swimmer, and Ant to investigate the speedup and solution quality of ASRacos.

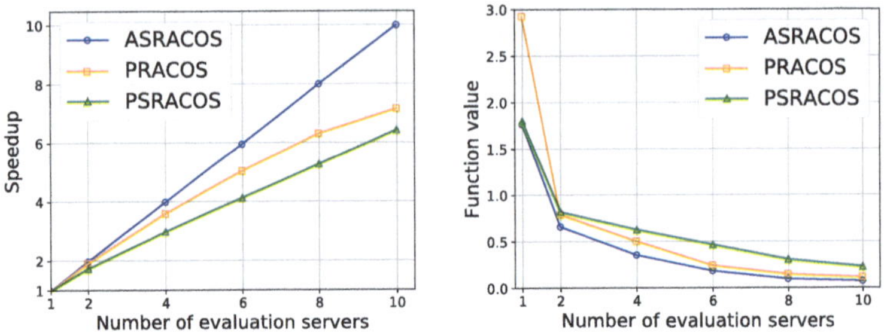

(a) On Sphere

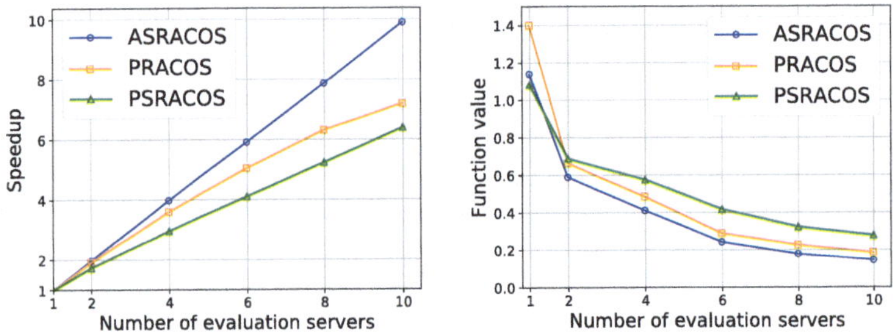

(b) On Ackley

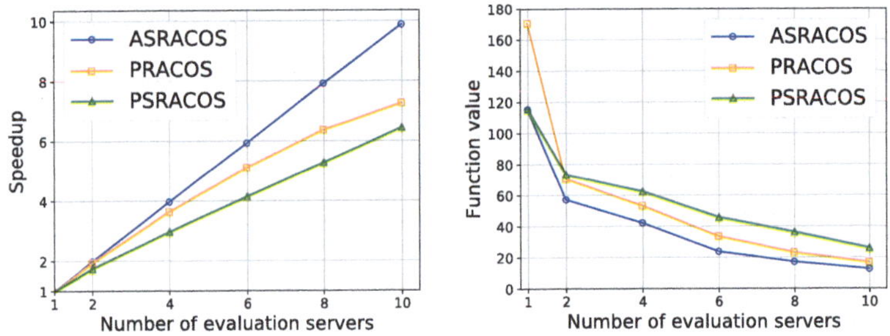

(c) On Rastrigin

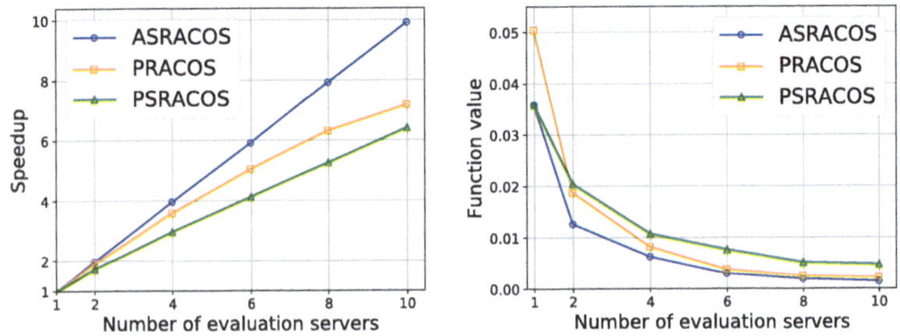

(d) On Griewank

Fig. 9.3 On each objective function, left: speedup, right: the average of the function value (the one closer to 0 the better) [4]

We use the framework of direct policy search to solve these tasks. Direct policy search employs optimization algorithms to search in the parameter space of a policy for maximizing the cumulative reward. The policy is often represented by a neural network [1, 2], whose weights $\boldsymbol{w} = \{w_1, w_2, \ldots, w_n\}$ are the parameters to be optimized. The neural network takes the observation of the state as input and outputs an action according to its policy. After that, it will get the reward of that action and the observation of the next state. This interaction can be repeated until the game is over or the maximum step is reached. The cumulative reward is used as an evaluation of the policy network, i.e., $f(\boldsymbol{w})_i = \sum_{t=1}^{T} R_t$. The agent would have different cumulative rewards if the initial state is reset to be different, so we take the average of multiple simulations as the final evaluation value of one neural network: $f(\boldsymbol{w}) = \sum_{i=1}^{m} f(\boldsymbol{w})_i/m$, which can reduce the noise to some extent. In a nutshell, our aim is to find the optimal parameter w for this network so as to achieve the best performance. We list the task information and the settings of neural network in Table 9.1, where d_{State}, #Actions, NN nodes, #Weights and Horizon respectively denote the dimension size of observation, the dimension size of action, the hidden layers of the neural network, the total number of parameters in the neural network and the maximum step.

We will briefly summarize each task and the details can be found in the homepage of OpenAI Gym. The Acrobot system includes two joints and two links, where the joint between the two links is actuated. Initially, the links are hanging downwards and the goal of this task is to swing the end of the low link up to a given height. In MountainCar, a car is on a one-dimensional track, positioned between two mountains. The goal is to drive up the mountain on the right through driving back and forth to build up momentum. In Pendulum, a pendulum starts in a random position, and the goal is to swing it up so it stays upright. HalfCheetah, Swimmer, and Ant are simulation tasks. In those tasks, a simulated object is controlled by a policy to achieve a specific goal. For example, in Ant, the policy should control a four-legged creature to make it walk forward as fast as possible. Among these tasks, Acrobot and MountainCar are finding policies with the smallest step number to achieve the goal. Other tasks are to find policies to get score from the environment as high as possible. The average cumulative reward of 200 simulations is used as the evaluation value of one network for Acrobot, MountainCar, and Pendulum. And for other tasks, the average reward

Table 9.1 Parameters of the Gym tasks [4]

Task	d_{State}	#Actions	NN nodes	#Weights	Horizon
Acrobot-v1	6	1	5, 3	48	500
MountainCar-v0	2	1	5	15	200
Pendulum-v0	3	1	5	20	200
HalfCheetah-v2	17	6	10	230	1000
Swimmer-v2	8	2	5, 3	61	1000
Ant-v2	111	8	15	1785	1000

of 20 simulations is used. The solution space X is set to be $[-10, 10]^{\#\text{Weight}}$. The output of the neural network is scaled to be within the action space, which is defined by the environment. Each algorithm is repeated 10 times and the mean value of the top-5 results is reported. The results are plotted in Figs. 9.4 and 9.5.

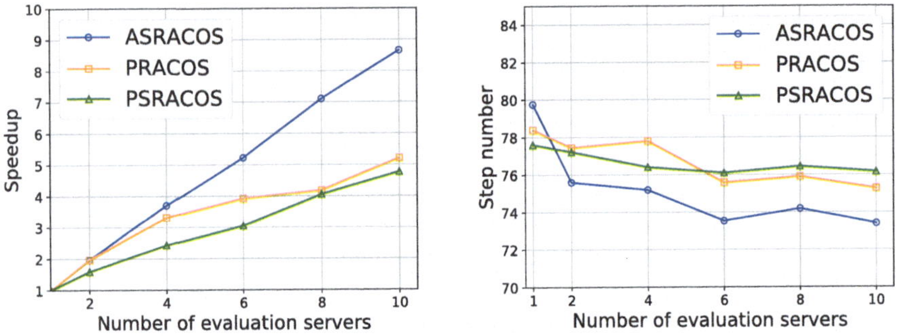

(a) On Acrobot

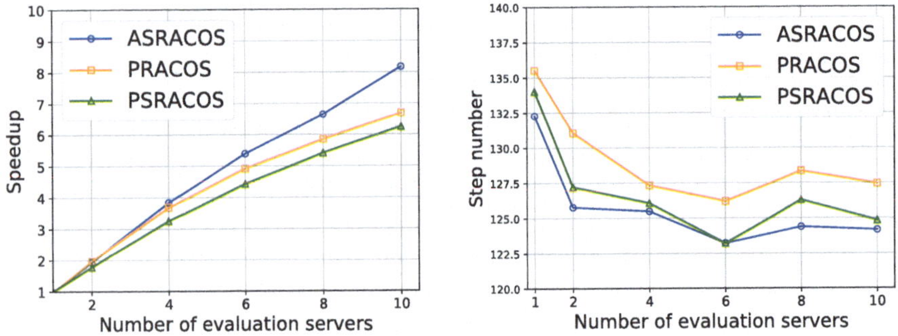

(b) On Mountain Car

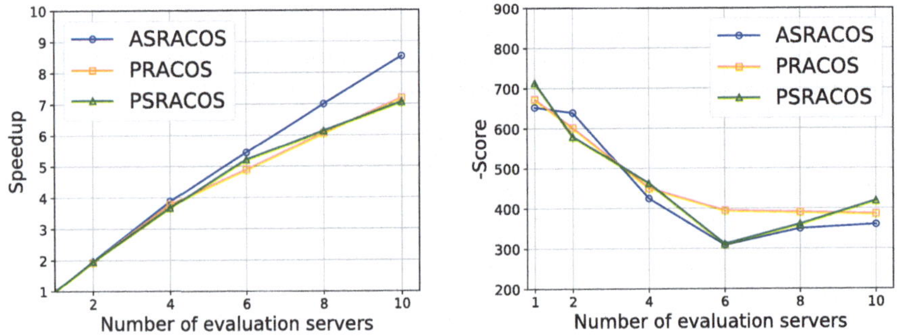

(c) On Pendulum

Fig. 9.4 For each task, left: speedup, right: the mean step (Acrobot, MountainCar) or minus score (Pendulum, HalfCheetah, Swimmer, Ant) of the best found policy (the smaller y-axis coordinate value the better) [4]

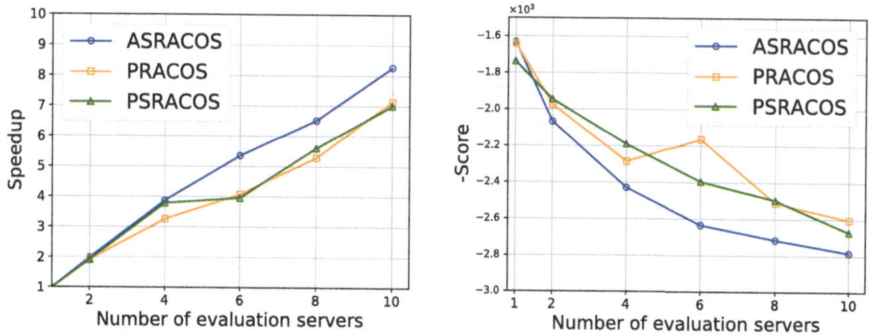

(a) On HalfCheetah

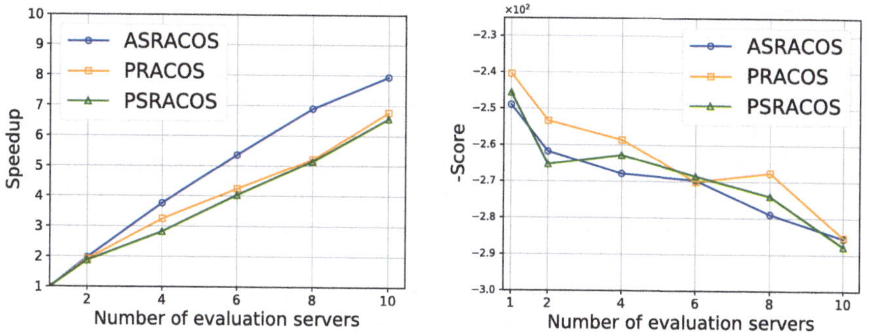

(b) On Swimmer

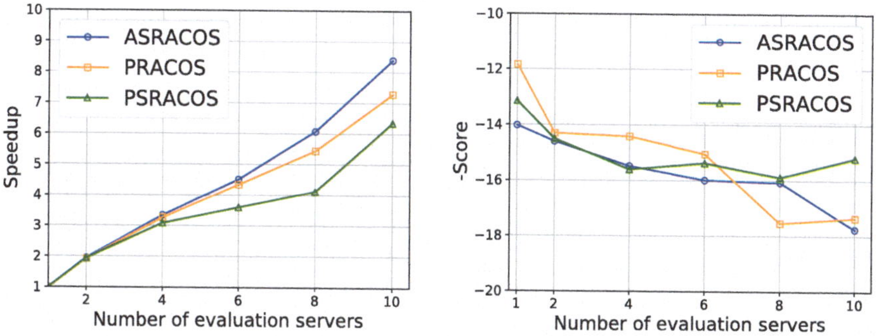

(c) On Ant

Fig. 9.5 For each task, left: speedup, right: the mean step (Acrobot, MountainCar) or minus score (Pendulum, HalfCheetah, Swimmer, Ant) of the best found policy (the smaller y-axis coordinate value the better) [4]

On Speedup
Budget is set to be 2000 for each algorithm. From the left plots of each task, we can observe that ASRACOS (blue line) can still achieve almost linear speedup, better than PRACOS and PSRACOS. Due to the competition for computing resource, the speedup ratio in these environments is smaller than that on synthetic functions, which simulate the time-consuming tasks simply by adding sleep operations. In addition, for Acrobot, MountainCar, and Ant, a better solution would make the game stop earlier, which consumes less evaluation time, and result in a lower speedup.

On Solution Value
We convert the maximization problems in Pendulum, HalfCheetah, Swimmer, and Ant to the minimization problems by adding a minus to the score. The time for optimization is set to be 20 min for each algorithm. From the right plots in each subfigure, we can see that the algorithm using more servers can get better solution quality in most cases. Nevertheless, in some cases, the algorithm may get worse solution quality. The reason is that in one case there exists randomness in the process of optimization, in another the evaluation is inaccurate under noisy environments, which may make a bad solution seem to be good and lead the optimization to the wrong direction. Similar to the results of the synthetic functions, ASRACOS achieves the best performance in most cases.

9.4 Summary

In this chapter, we present an asynchronous derivative-free classification-based optimization method, ASRACOS, originally proposed in [4], for accelerating the optimization. We analyze the query complexity of ASRACOS and further provide the condition on which ASRACOS can achieve a better (worse) performance than SRACOS using the same number of evaluations. In experiments, we first study the convergence rate of ASRACOS on synthetic functions, showing that ASRACOS can achieve higher convergence rate when having more evaluation servers. On both synthetic functions and direct policy search for controlling tasks, ASRACOS demonstrates almost linear speedup and gets a better solution quality than other parallel algorithms, which verifies the effectiveness of asynchronous parallelism. Future work includes combining noise-handling methods into ASRACOS to speed up the optimization under noisy environments and applying ASRACOS to large-scale optimization problems in real world.

References

1. El-Fakdi A, Carreras M, Palomeras N (2005) Direct policy search reinforcement learning for robot control. In: Proceedings of the 8th international conference of the ACIA artificial intelligence research and development, Alguer, Italy, pp 9–16
2. El-Fakdi A, Carreras M, Palomeras N (2006) Towards direct policy search reinforcement learning for robot control. In: Proceedings of the 2006 IEEE/RSJ international conference on intelligent robots and systems, Beijing, China, pp 3178–3183
3. Hu YQ, Qian H, Yu Y (2017) Sequential classification-based optimization for direct policy search. In: Proceedings of the 31st AAAI conference on artificial intelligence, San Francisco, CA, pp 2029–2035
4. Hu YQ, Qian H, Yu Y (2019) Asynchronous classification-based optimization. In: Proceedings of the 1st international conference on distributed artificial intelligence, Beijing, China, pp 9:1-9:8.https://doi.org/10.1145/3356464.3357709
5. Qian C, Shi JC, Yu Y, Tang K, Zhou ZH (2016) Parallel Pareto optimization for subset selection. In: Proceedings of the 25th international joint conference on artificial intelligence, New York, NY, pp 1939–1945
6. Yu Y, Qian H, Hu YQ (2016) Derivative-free optimization via classification. In: Proceedings of the 30th AAAI conference on artificial intelligence, Phoenix, Arizona, pp 2286–2292
7. Yu Y, Qu WY, Li N, Guo Z (2017) Open category classification by adversarial sample generation. In: Proceedings of the 26th international joint conference on artificial intelligence, Melbourne, Australia, pp 3357–3363
8. Zhang J, Sun Y, Huang S, Nguyen CT, Wang X, Dai X, Chen J, Yu Y (2017) AGRA: an analysis-generation-ranking framework for automatic abbreviation from paper titles. In: Proceedings of the 26th international joint conference on artificial intelligence, Melbourne, Australia, pp 4221–4227
9. Zhou WJ, Yu Y, Zhang ML (2017) Binary linear compression for multi-label classification. In: Proceedings of the 26th international joint conference on artificial intelligence, Melbourne, Australia, pp 3546–3552
10. Zinkevich M, Weimer M, Smola A, Li L (2010) Parallelized stochastic gradient descent. In: Advances in neural information processing systems, vol 23, British Columbia, Canada, pp 2595–2603

Chapter 10
Toolbox: ZOOpt

Abstract This chapter introduces the ZOOpt toolbox, a powerful tool for zeroth-order optimization designed to address high-dimensional and noisy optimization problems, particularly in machine learning tasks such as hyper-parameter tuning and direct policy search. ZOOpt implements state-of-the-art algorithms, including SRacos, ASRacos, and POSS, and supports optimization in continuous, discrete, and hybrid spaces. It also features noise-handling mechanisms like value suppression and threshold selection, as well as high-dimensionality handling through sequential random embedding. The toolbox integrates with the Ray framework for distributed optimization, enabling efficient parallel computation. Empirical studies demonstrate ZOOpt's superior convergence rate, scalability, and robustness against noise compared to other optimization toolboxes. Experiments on synthetic functions and machine learning tasks, including classification with Ramp loss and OpenAI Gym control tasks, highlight ZOOpt's effectiveness. The chapter concludes with a summary of ZOOpt's capabilities and its potential for real-world applications.

This chapter introduces the ZOOpt (**Z**eroth **O**rder **Opt**imization) toolbox [7], which provides major algorithms introduced in the previous chapters. ZOOpt implements single-machine parallel optimization using Python and multi-machine distributed optimization for time-consuming tasks by incorporating the Ray framework, a popular platform for building distributed applications. ZOOpt particularly focuses on optimization problems in machine learning, addressing high-dimensional and noisy problems such as hyper-parameter tuning and direct policy search. The toolbox is maintained as a ready-to-use tool for real-world machine learning tasks.

10.1 Methods in ZOOpt

In Table 10.1, we summarize the algorithms implemented in the ZOOpt toolbox, along with their support for different search spaces, parallelization, and compatibility with noise and high-dimensional handlers.

Optimization in continuous/discrete/hybrid spaces. ZOOpt implements SRACOS (Chap. 6 as the default optimization method, which has shown high efficiency

Y. Yu et al., *Derivative-Free Optimization*, Machine Learning: Foundations, Methodologies, and Applications, https://doi.org/10.1007/978-981-96-5929-6_10

Table 10.1 Algorithms implemented in the ZOOpt toolbox [7]. For each algorithm, we conclude its support on different kinds of search space, parallelization and the compatibility with the noise handler and the high-dimensional handler

Algorithms in ZOOpt		Search space			Parallelization			Noise Handler	High-dimensionality handler	Suitable tasks
		Continuous	Discrete	Hybrid	Single-machine	Multi-machine	Asynchronous			
Classification-based optimization	Racos	✓	✓	✓	✓			✓	✓	Non-differential, non-convex, noisy and high-dimensional functions
	SRacos	✓	✓	✓	✓			✓	✓	
	ASRacos	✓	✓	✓	✓	✓	✓	✓	✓	
Pareto optimization for subset selection	POSS		✓					✓		Subset selection problems
	PPOSS		✓		✓			✓		

in a range of learning tasks. Optional methods are RACOS (Chap. 5) and ASRACOS (Chap. 9), which are the batch and asynchronous versions of SRACOS, respectively. A routine is in place to set up the default parameters of the two methods, while users can override them. Benefiting from the compatibility of the classifier with multiple data types, classification-based optimization naturally supports optimization in continuous, discrete (categorical), or hybrid spaces.

Optimization in binary vector space with constraints. If the optimization task is in a binary vector space with constraints, such as the subset selection problem, POSS [9] is the default optimization method. POSS treats subset selection tasks as a bi-objective optimization problem that simultaneously optimizes a given criterion and the subset size. POSS has been proven to have the best-so-far approximation quality on these problems. PPOSS [10] is the parallel version of the POSS algorithm.

Noise handling. Noise has a great impact on the performance of derivative-free optimization. Resampling is the most straightforward method to handle noise, which evaluates one sample several times to obtain a stable mean value. Besides resampling, more efficient methods, including value suppression (Chap. 8) and threshold selection [11], are implemented in ZOOpt.

High-dimensionality handling. An increase in the search space dimensionality badly injures the performance of derivative-free optimization. When a high-dimensional search space has a low effective dimension, random embedding [12] is an effective way to improve efficiency. Also, sequential random embeddings (Chap. 7) can be used when there is no clear low effective dimension.

Distributed optimization. Evaluation of a sampled solution is usually time-consuming for many real-world optimization tasks, such as hyper-parameter tuning in large-scale machine learning projects. Incorporating the Ray framework [8], ZOOpt implements an efficient distributed optimization module that enables users to parallelize single-machine code with little to no code changes.

10.2 Usage

This section briefly introduces single-machine optimization, distributed optimization, optimization under noise, and optimization in high-dimensional spaces through a few examples. For the full tutorial, including detailed API introduction, hyper-parameter tuning tricks, and all examples, we refer readers to https://zoopt.readthedocs.io/en/latest/.

Single-machine optimization. The core architecture of ZOOpt includes three parts: *Objective*, *Parameter*, and *Opt.min*. The *Objective* object defines the function expression and the search space. The *Parameter* object defines all parameters used by the optimization algorithm. *Opt.min* is the interface for performing optimization. After defining a user-specified objective function and the corresponding search space, only one line of code is needed to perform optimization using *Opt.min*. A quick-start example is provided as follows.

```
import numpy as np
from zoopt import ValueType, Dimension2, Objective,
                              Parameter, Opt

def ackley(solution):
        x = solution.get_x()
        bias = 0.2
        value = -20 * np.exp(-0.2 * np.sqrt(sum([(i
                                             - bias) * (i -
                                             bias) for i in x]
                                             ) / len(x))) - \
                np.exp(sum([np.cos(2.0*np.pi*(i-bias
                                                      )) for i
                                                      in x]) /
                                                      len(x)) +
                                                       20.0 +
                                                      np.e
        return value

dim_size = 100  # dimension size
dim = Dimension2([(ValueType.CONTINUOUS, [-1, 1], 1e
                              -6)]*dim_size)
obj = Objective(ackley, dim)
# perform optimization
solution = Opt.min(obj, Parameter(budget=100*
                              dim_size))
# print the solution
print(solution.get_x(), solution.get_value())
# parallel optimization for time-consuming tasks
solution = Opt.min(obj, Parameter(budget=100*
                              dim_size, parallel=True,
                              server_num=3))
```

Distributed optimization. Distributed optimization in ZOOpt is implemented by incorporating Ray. Currently, ZOOpt is an optional optimization tool in *Ray.tune*, a library for fast hyper-parameter tuning at any scale. Through *Ray.tune*, users can easily distribute the optimization without worrying about the communication infrastructure. We provide an example as follows:

```
import time
from ray import tune
from ray.tune.suggest.zoopt import ZOOptSearch
from ray.tune.schedulers import
                              AsyncHyperBandScheduler
from zoopt import ValueType  # noqa: F401

def evaluation_fn(step, width, height):
    time.sleep(0.1)
    return (0.1 + width * step / 100)**(-1) + height
                                          * 0.1

def easy_objective(config):
```

```
    # Hyperparameters
    width, height = config["width"], config["height"
                                ]

    for step in range(config["steps"]):
        # Iterative training function - can be any
                                        arbitrary
                                        training
                                        procedure
        intermediate_score = evaluation_fn(step,
                                        width, height)
        # Feed the score back back to Tune.
        tune.report(iterations=step, mean_loss=
                                        intermediate_score
                                        )

if __name__ == "__main__":
    import argparse
    parser = argparse.ArgumentParser()
    parser.add_argument(
        "--smoke-test", action="store_true", help="
                                        Finish quickly
                                        for testing")
    parser.add_argument(
        "--server-address",
        type=str,
        default=None,
        required=False,
        help="The address of server to connect to if
                                        using "
        "Ray Client.")
    args, _ = parser.parse_known_args()

    if args.server_address:
        import ray
        ray.init(f"ray://{args.server_address}")
    num_samples = 10 if args.smoke_test else 1000
    zoopt_search_config = {
        "parallel_num": 8,
    }
    zoopt_search = ZOOptSearch(
        algo="Asracos",  # only support ASRacos
                                        currently
        budget=num_samples,
        **zoopt_search_config)
    scheduler = AsyncHyperBandScheduler()
    analysis = tune.run(
        easy_objective,
        metric="mean_loss",
        mode="min",
        search_alg=zoopt_search,
        name="zoopt_search",
        scheduler=scheduler,
```

```
        num_samples=num_samples,
        config={
            "steps": 10,
            "height": tune.quniform(-10, 10, 1e-2),
            "width": tune.randint(0, 10)
        })
    print("Best config found: ", analysis.
                                    best_config)
```

Optimization under noise. The noise handler can be enabled by adding some attributes to the definition of the *Parameter* object. Three kinds of noise handlers are implemented in ZOOpt: Naive resampling, value suppression, and threshold selection. Naive resampling reduces noise by evaluating the same solution multiple times and taking their mean value as the final result. Value suppression reduces noise more efficiently by re-evaluating the best solution when it isn't updated for a pre-defined number of times. Threshold selection is a noise handler customized for the POSS algorithm, where solution x is better than y only if $f(x)$ is smaller than $f(y)$ by at least a threshold. We provide simplified cases on how to use these noise handlers as follows. Their full versions can be found in the tutorial.

```
from zoopt import Parameter
from sparse_mse import SparseMSE
import numpy as np

# naive resampling
parameter = Parameter(budget=200000, noise_handling=
                        True, resampling=True,
                        resample_times=10)
# value suppression
parameter = Parameter(budget=200000, noise_handling=
                        True, suppression=True,
                        non_update_allowed=500,
                        resample_times=100,
                        balance_rate=0.5)
# threshold selection
mse = SparseMSE('sonar.arff')
mse.set_sparsity(8)
parameter = Parameter(algorithm='poss',
                        noise_handling=True,
                        ponss=True, ponss_theta=0
                        .5, ponss_b=mse.get_k(),
                        budget=2 * np.exp(1) * (
                        mse.get_sparsity() ** 2)
                        * mse.get_dim().get_size
                        ())
```

Optimization in high-dimensional spaces. ZOOpt contains the sequential random embedding (SRE) (Chap. 7) to handle high-dimensional optimization problems. SRE runs the optimization algorithms in a low-dimensional space, where the function values of solutions are evaluated via embedding into the original high-dimensional space sequentially. SRE is effective for the function class where all dimensions may affect the function value, but many of them only have a small bounded effect, and can scale RACOS, SRACOS, and ASRACOS (the main optimization algorithms in ZOOpt) to 100,000-dimensional problems. The high-dimensionality handler can be enabled by adding attributes to the definition of the *Parameter* object. An example is provided as follows:

```
from simple_function import sphere_sre
from zoopt import Dimension, ValueType, Dimension2,
                              Objective, Parameter,
                              ExpOpt
dim_size = 10000  # dimension size
dim_regs = [[-1, 1]] * dim_size  # search space
dim_tys = [True] * dim_size  # continuous
dim = Dimension(dim_size, dim_regs, dim_tys)  # form up
                              the dimension object
objective = Objective(sphere_sre, dim)  # form up the
                              objective function
budget = 2000  # number of calls to the objective
                              function
parameter = Parameter(budget=budget,
                              high_dimensionality_handling
                              =True, reducedim=True,
                              num_sre=5, low_dimension=
                              Dimension(10, [[-1, 1]] *
                              10, [True] * 10))
solution_list = ExpOpt.min(objective, parameter, repeat=
                              1, plot=True)
```

10.3 Experiments

In our experiments, we aim to answer the following questions: (1) How does ZOOpt compare to prior derivative-free optimization toolboxes on classic optimization benchmarks? (2) Can ZOOpt scale better than other toolboxes when the dimension size of the optimization task increases? (3) Can ZOOpt have better robustness against noise than other toolboxes? (4) How does ZOOpt compare to other toolboxes on machine learning tasks?

To answer these questions, we compare ZOOpt to several prior derivative-free optimization toolboxes, including pycma,[1] DEAP,[2] pygad,[3] and Hyperopt.[4] Pycma [6] is a Python implementation of the CMA-ES [5] algorithm. DEAP [3] is an evolutionary computation framework. Pygad [4] is an open-source Python library of genetic algorithms. Hyperopt [1] implements state-of-the-art Bayesian optimization

[1] https://github.com/CMA-ES/pycma.

[2] https://github.com/DEAP/deap.

[3] https://github.com/ahmedfgad/GeneticAlgorithmPython.

[4] https://github.com/hyperopt/hyperopt.

algorithms for hyper-parameter tuning. For all toolboxes, we choose the default algorithm and the recommended parameters according to their tutorials. It's worth noting that each toolbox actually implements many optimization algorithms. However, we don't exhaust the algorithm-level comparisons in this paper. Instead, we choose the default algorithm and focus more on the toolbox itself. The source code of the experiments can be found at https://github.com/AlexLiuyuren/ZOOpt_experiment.

Experiments are conducted on three kinds of tasks. To answer questions (1), (2), and (3), we conduct experiments on optimizing benchmark synthetic functions. We empirically evaluate the performance of the tested toolboxes, including the convergence rate, scalability, and robustness against noise, on four benchmark synthetic functions. To answer question (4), we conduct experiments on two machine learning tasks. We study a classification task with Ramploss, where the objective function is similar to that of support vector machines (SVM), but the loss function is the Ramp loss instead of the hinge loss used in SVM. We then study direct policy search for controlling tasks, where the policy is a fully connected feedforward neural network, and its weights are optimized directly by derivative-free optimization algorithms.

10.3.1 Results on Optimizing Synthetic Functions

To answer questions (1), (2), and (3), we conduct experiments on optimizing benchmark synthetic functions. Among them, the Ackley, Rastrigin, and Schwefel functions are highly non-convex, while the Sphere function is convex. The optimal values of the four functions are all zero. The Ackley and Sphere functions are minimized within the search space $X = [-1, 1]^d$, where d is the dimension size. The Rastrigin function is minimized within $[-5, 5]^d$. The Schwefel function is minimized within $[-500, 500]^d$. The optimal position of each function (except the Schwefel function, which is fixed to [420.97, ..., 420.97]) is shifted from [0, ..., 0] to a random point sampled from $[0.2 * l, 0.2 * u]^d$, where l and u respectively refer to the lower and upper bounds of the search space on that dimension. This is to avoid a possible optimization bias toward the origin point. The 3-d graphs of these functions are shown in Fig. 10.1. Each experiment is repeated 30 times. Mean values and 95% confidence intervals are recorded. Results are shown in Fig. 10.2.

Convergence rate. We set the dimension size to be 20 for each objective function and the number of evaluations to be 2000. We study the convergence rate with regard to the number of function evaluations by recording the best-so-far solution value during the optimization. As shown in Fig. 10.2, ZOOpt reduces the objective function value at the highest rate in all tasks.

Scalability. The scalability of derivative-free optimization methods is critical for solving large-scale problems. In this experiment, we quantitatively study the scalability of ZOOpt. We set the dimension size d to be 20, 200, 400, 600, 800, 1000 and the number of function evaluations to be $100 \times d$. The confidence interval is omitted for clarity. Figure 10.3 shows that ZOOpt has the lowest growth rate of the

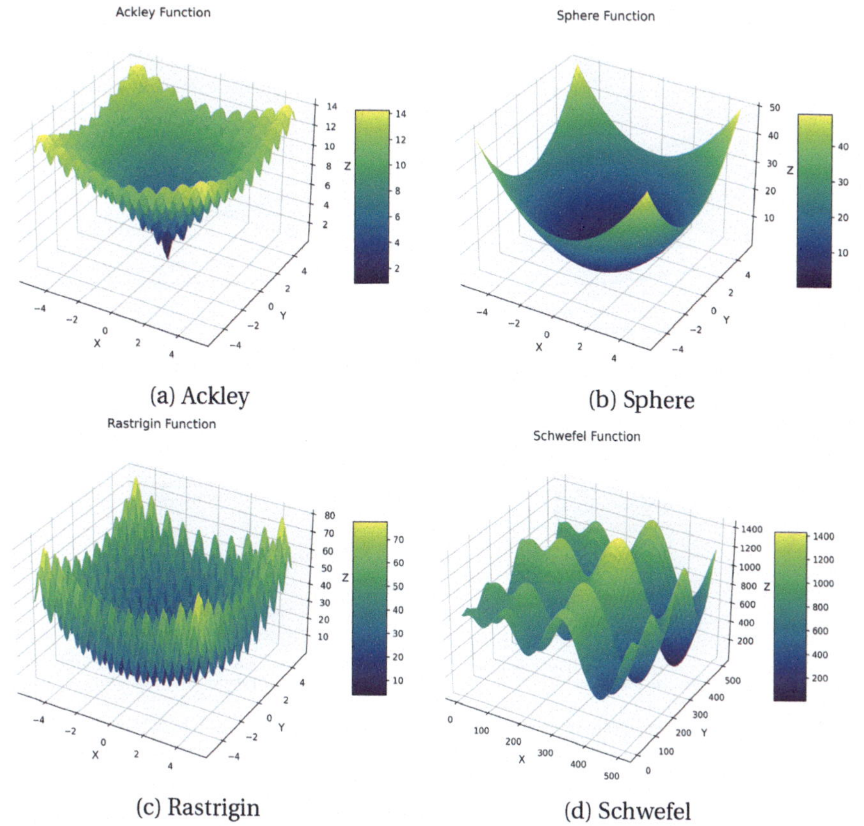

(a) Ackley (b) Sphere (c) Rastrigin (d) Schwefel

Fig. 10.1 3-d graphs of four benchmark synthetic functions. Among them, the Ackley, Rastrigin, and Schwefel functions are highly non-convex, while the Sphere function is convex

function value in all tasks as the dimension size increases, indicating that ZOOpt has better scalability than other toolboxes.

Robustness against noise. To study the performance of ZOOpt on optimizing noisy objectives, we add Gaussian noise to the original functions to simulate the noisy environment. The new objective functions are defined as $f^N(x) = f(x) + N(0, \sigma^2)$. The number of function evaluations is set to 10000. For all tasks, ZOOpt and pycma use their built-in noise handlers, while DEAP and pygad do not. Figure 10.4 shows that ZOOpt reduces the function value at a steady pace as the number of evaluations increases, despite the noise.

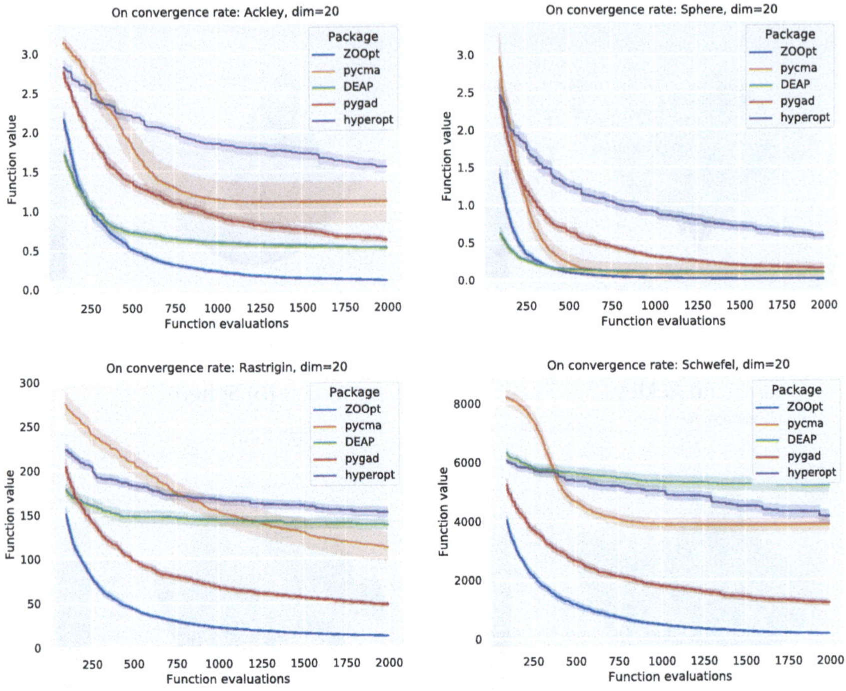

Fig. 10.2 The convergence rate of the tested toolboxes on four minimization benchmark synthetic functions [7]

10.3.2 Results on Classification Tasks with Ramploss

The Ramp loss is defined as $R_s(z) = H_1(z) - H_s(z)$ with $s < 1$, where $H_s(z) = \max\{0, s - z\}$ is the Hinge loss with s being the Hinge point. The task is to find a vector w and a scalar b to minimize $f(w, b) = \frac{1}{2}\|w\|_2^2 + C\sum_{\ell}^{L} R_s(y_\ell(w^\top v_\ell + b))$, where v_l is the training instance and $y_l \in \{-1, +1\}$ is its label. Due to the convexity of the Hinge loss, the number of support vectors increases linearly with the number of training instances in SVM, which is undesirable with respect to scalability. This problem can be alleviated by using the Ramp loss [2].

We employ two binary class UCI datasets, Adult and Bank, for the classification task. Discrete variables of the original features are preprocessed by one-hot encoding. Continuous variables are normalized into $[-1, 1]$. The resulting feature dimension (excluding the label) is expanded to 108 for Adult and 51 for Bank. Since we focus on the optimization performance, we only compare the results on the complete dataset. Two hyper-parameters, i.e., C and s, are adjustable in the optimization formulation. We set $s \in \{-1, 0\}$ and $C \in \{0.1, 0.5, 1, 2, 5, 10\}$ to study the effectiveness of the tested toolboxes under different hyper-parameter settings. We set the total number

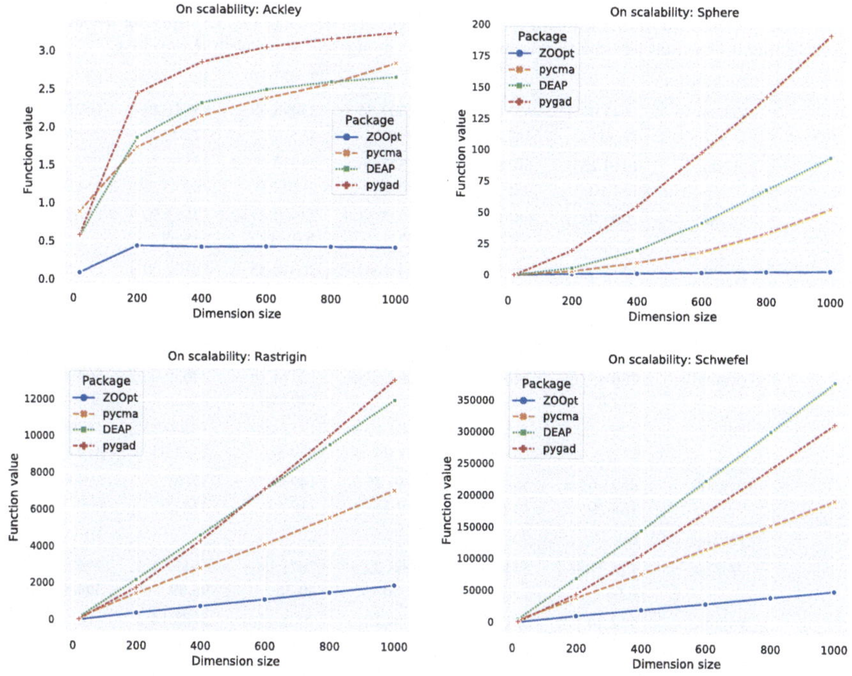

Fig. 10.3 The scalability of the tested toolboxes as the dimension size increases [7]

of calls to the objective function to be $40n$ for all toolboxes, where n is the number of instances. The achieved objective values are reported in Table 10.2.

It can be observed that ZOOpt is comparable with pycma and dominates DEAP and pygad in all cases. Notice that the smaller C is, the closer the objective function is to convexity. Therefore, the optimization difficulty increases with C. Although the results of ZOOpt and pycma are close, ZOOpt achieves better results when C is large, i.e., when the objective function is further from convexity. Pycma is better when the objective function is closer to convexity.

10.3.3 Results on Direct Policy Search for OpenAI Controlling Tasks

Gym tasks. In the OpenAI Gym environment, we use six existing controlling tasks: 'Acrobot', 'MountainCar', 'HalfCheetah', 'Hopper', 'Humanoid', and 'Swimmer', to test the toolboxes. We apply a feedforward neural network as the policy. The task information and neural network structures are shown in Table 10.3. For example, in 'Acrobot': $|S| = 6$, $|A| = 3$; the neural network has two hidden layers with 5 and 3 neurons each; $|w| = 48$; the activation functions for hidden layers and the output

Table 10.2 Results on the Adult (upper) and Bank (lower) datasets [7]. Comparing the achieved objective function values against the parameter C of the classification with Ramp loss

S	Package \ C	0.1	0.5	1	2	5	10
−1	ZOOpt	1642.07 ± 79.33	6331.05 ± 147.10	12002.56 ± 287.74	**23098.68** ± 435.40	**55151.49** ± 772.15	**108896.69** ± 1944.12
	pycma	**1414.25** ± 154.10	**6028.83** ± 495.39	**11537.06** ± 120.91	23259.40 ± 2184.85	55576.41± 762.45	109422.90± 944.75
	DEAP	2005.05 ± 88.32	6822.13 ± 157.85	12625.33 ± 257.62	23909.87 ± 303.31	57152.50 ± 845.67	111093.63 ± 1454.88
	pygad	3315.09 ± 146.83	8643.11 ± 276.07	14456.62 ± 240.14	26048.55 ± 381.98	59147.50 ± 440.53	113461.05 ± 840.28
0	ZOOpt	1001.56 ± 29.79	3585.84 ± 160.28	**6665.13** ± 408.27	**12451.74** ± 247.92	**29583.38** ± 1886.19	57042.13 ± 751.92
	pycma	**780.45** ± 32.60	**3406.22** ± 345.54	6668.67 ± 776.18	12715.15 ± 1493.09	29639.06 ± 2585.68	**56650.23** ± 509.49
	DEAP	1297.46 ± 41.37	4159.68 ± 201.96	7185.12 ± 457.99	13124.33 ± 859.58	30400.34 ± 1767.03	58898.11 ± 3797.57
	pygad	2531.69 ± 164.29	5588.36 ± 146.79	8846.75 ± 300.44	14949.72 ± 710.44	32167.27 ± 474.29	60436.24 ± 600.72
S	Package \ C	0.1	0.5	1	2	5	10
−1	ZOOpt	128.31 ± 6.69	545.45 ± 7.45	1068.09 ± 12.00	**2075.12** ± **40.36**	**5045.72** ± **98.89**	**9957.51** ± **306.85**
	pycma	**114.24** ± **5.82**	**531.11** ± **4.59**	**1056.25** ± **6.63**	2088.15 ± 30.01	5185.14 ± 89.46	110236.24 ± 280.79
	DEAP	248.58 ± 22.72	670.73 ± 21.44	1191.32 ± 24.96	2234.39 ± 19.57	5307.23 ± 67.89	10316.18 ± 226.76
	pygad	627.27 ± 69.33	1055.97 ± 61.62	1564.35 ± 77.59	2618.18 ± 66.00	5753.68 ± 86.63	10893.56 ± 119.12
0	ZOOpt	73.69 ± 6.61	285.04 ± 9.02	545.82 ± 4.82	1064.49 ± 6.32	**2618.38** ± **52.69**	**5091.75** ± **124.31**
	pycma	**60.84** ± **4.08**	**270.39** ± **3.49**	**532.24** ± **5.70**	**1053.18** ± **3.72**	2620.21 ± 11.10	5221.21 ± 31.35
	DEAP	192.68 ± 16.94	415.67 ± 24.26	673.14 ± 21.71	1187.59 ± 16.59	2763.42 ± 17.70	5329.71 ± 44.41
	pygad	543.22 ± 60.52	798.79 ± 73.10	1037.72 ± 81.79	1573.16 ± 89.69	3145.45 ± 80.23	5787.98 ± 71.72

layer are ReLU and softmax, respectively; the maximum number of steps is 500. We will give a summary of each task. More details can be found on the homepage of OpenAI Gym. In ‘Acrobot’, the system includes two joints and two links, where the joint between the two links is actuated. Initially, the links are hanging downwards, and the goal of this task is to swing the end of the low link up to a given height. In ‘MountainCar’, a car is positioned in a valley between two mountains and wants to drive up the mountain on the right by building up momentum. ‘HalfCheetah’, ‘Hopper’, ‘Humanoid’, and ‘Swimmer’ are simulation tasks. In those tasks, the policy controls simulated objects to achieve a goal. For example, in ‘HalfCheetah’, the policy should control a cheetah with half body to run forward as fast as possible. The tasks of ‘Acrobot’ and ‘MountainCar’ are finding policies with the smallest step

Table 10.3 The parameters of the direct policy search for OpenAI controlling tasks [7]

Task name	d_{State}	Action type	Action size	NN nodes	#Weights	Activation (hidden)	Activation (output)	Horizon
Acrobot-v1	6	Discrete	3	5, 3	54	relu	softmax	500
MountainCar-v0	2	Discrete	3	5	25	relu	softmax	200
HalfCheetah-v2	17	Continuous	6	10	230	relu	tanh	1000
Hopper-v2	11	Continuous	3	9,5	159	relu	tanh	1000
Humanoid-v2	376	Continuous	17	25	9825	relu	tanh	1000
Swimmer-v2	8	Continuous	2	5,3	61	relu	tanh	1000

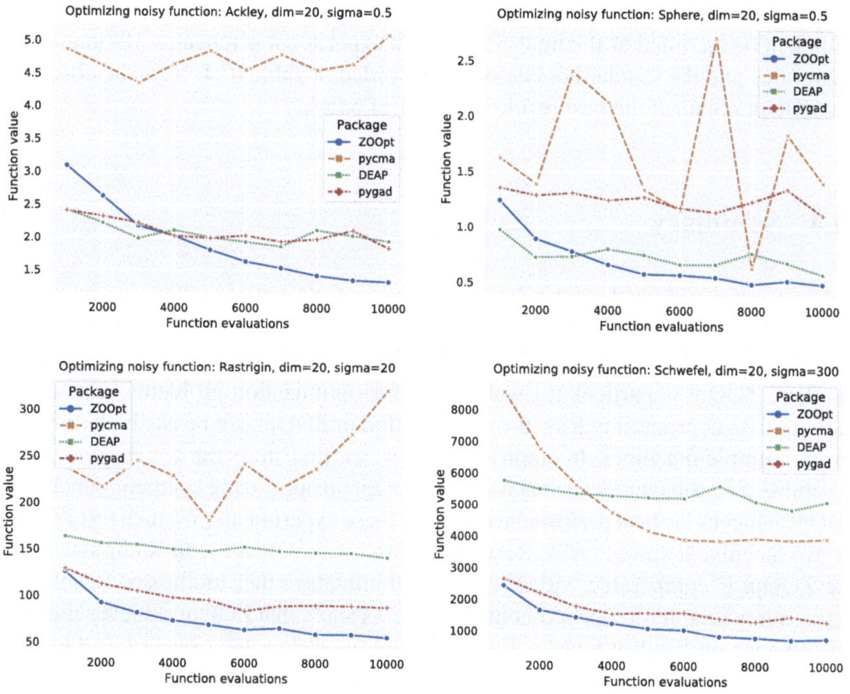

Fig. 10.4 The performance on optimizing noisy functions [7]

number when goals are met. The tasks except for 'Acrobot' and 'MountainCar' are finding policies to control objects to get scores from the environment as high as possible. Therefore, in Table 10.4, the columns of 'Acrobot' and 'MountainCar' are step numbers, where smaller is better. The other rows are the cumulative rewards from environments, where larger is better.

The average cumulative rewards of 10 simulations are used as the evaluation value of a neural network to reduce noise. The solution space X is set to be $[-10, 10]^{\#\text{Weight}}$. The output of the neural network is scaled to be within the action space, which is defined by the environment. All toolboxes use 2,000 evaluations for each task. The

Table 10.4 The mean scores and the standard deviation of the best found policy by each toolbox [7]. The numbers in bold represent the best scores in each column. The down arrow means the score is better when smaller, and the up arrow means better when larger

Package	Acrobot-v1 ↓	MountainCar-v0 ↓	HalfCheetah-v2 ↑	Hopper-v2 ↑	Humanoid-v2 ↑	Swimmer-v2 ↑
ZOOpt	**82.02 ± 3.05**	**128.23 ± 12.41**	1295.39 ± 731.71	**738.86 ± 391.06**	**448.93 ± 80.33**	**138.05 ± 107.95**
pycma	314.40 ± 186.55	197.81 ± 6.56	465.50 ± 492.81	305.27 ± 358.43	398.30 ± 111.12	35.39 ± 32.06
DEAP	144.26 ± 121.82	200.00 ± 0.00	**1409.11 ± 437.10**	224.53 ± 259.22	303.29 ± 110.66	75.05 ± 104.25
pygad	207.66 ± 146.10	174.85 ± 33.98	188.32 ± 809.55	181.45 ± 230.35	293.49 ± 102.77	50.03 ± 102.21

best solution will be re-evaluated 30 times to further reduce the noise, and their mean value will be recorded as the final result. Each experiment is repeated 10 times. The mean value and the standard deviation are recorded in Table 10.4. It can be observed that ZOOpt obtained the best results on 5 out of 6 tasks.

10.4 Summary

In this chapter, we introduce the ZOOpt toolbox, which provides efficient derivative-free solvers and is designed to be easy to use. By combining several state-of-the-art classification-based optimization methods, noise handlers, and high-dimensionality handlers, ZOOpt is particularly well-suited for optimization problems in machine learning. By incorporating Ray, the optimization in ZOOpt can be easily distributed across multiple machines. In empirical studies, we first study the convergence rate, scalability, and robustness against noise of ZOOpt on optimizing synthetic functions. ZOOpt achieves the best performance in all of these experiments. We then test ZOOpt on two machine learning tasks. Results on classification tasks with Ramploss show that ZOOpt is comparable with pycma and dominates other toolboxes. Results on direct policy search for OpenAI controlling tasks show that ZOOpt achieves the best performance on 5 out of 6 tasks. For a detailed tutorial on the usage of ZOOpt, we refer readers to the project homepage https://github.com/polixir/ZOOpt.

References

1. Bergstra J, Yamins D, Cox D (2013) Making a science of model search: hyperparameter optimization in hundreds of dimensions for vision architectures. In: Proceedings of the 30th international conference on machine learning, vol 28, Atlanta, GA, pp 115–123
2. Collobert R, Sinz F, Weston J, Bottou L (2006) Trading convexity for scalability. In: Proceedings of the 23rd international conference on machine learning, vol 148, pp 201–208, Pittsburgh, Pennsylvania
3. Fortin F-A, De Rainville F-M, Gardner M-A, Parizeau M, Gagné C (2012) DEAP: evolutionary algorithms made easy. J Mach Learn Res 13:2171–2175
4. Gad AF (2021) PyGAD: an intuitive genetic algorithm Python library. CoRR, abs/2106.06158

5. Hansen N, Müller SD, Koumoutsakos P (2003) Reducing the time complexity of the derandomized evolution strategy with covariance matrix adaptation (CMA-ES). Evol Comput 11(1):1–18
6. Hansen N, Akimoto Y, Baudis P (2019). CMA-ES/pycma on Github Zenodo. https://doi.org/10.5281/zenodo.2559634
7. Liu Y-R, Hu Y-Q, Qian H, Yu Y, Qian C (2022) Zoopt: toolbox for derivative-free optimization. Sci China Inf Sci 65:207101
8. Moritz P, Nishihara R, Wang S, Tumanov A, Liaw R, Liang E, Elibol M, Yang Z, Paul W, Jordan MI, Stoica I (2018) Ray: A distributed framework for emerging AI applications. In: 13th USENIX symposium on operating systems design and implementation, Carlsbad, CA, pp 561–577
9. Qian C, Yu Y, Zhou ZH (2015) Subset selection by pareto optimization. In: Advances in neural information processing systems, vol 28, Montreal, Canada, pp 1765–1773
10. Qian C, Shi JC, Yu Y, Tang K, Zhou ZH (2016) Parallel pareto optimization for subset selection. In: Kambhampati S (ed) Proceedings of the 25th international joint conference on artificial intelligence, New York, NY. IJCAI/AAAI Press, pp 1939–1945
11. Qian C, Shi JC, Yu Y, Tang K, Zhou ZH (2017) Subset selection under noise. In: Advances in neural information processing systems, vol 30, Long Beach, CA, pp 3563–3573
12. Wang Z, Zoghi M, Hutter F, Matheson D, Freitas ND (2016) Bayesian optimization in a billion dimensions via random embeddings. J Artif Intell Res 55:361–387

Part IV
Application to Automatic Machine Learning

Chapter 11
Experienced Optimization: Acceleration in Hyper-Parameter Optimization

Abstract This chapter explores the concept of experienced optimization in hyper-parameter optimization, a critical task in Automatic Machine Learning (AutoML). Hyper-parameter optimization often involves derivative-free optimization (DFO) methods, which can be inefficient due to the high cost of evaluating hyper-parameter configurations. The chapter introduces an experienced optimization approach that leverages historical optimization data to improve efficiency in new tasks. Two algorithms, ExpSRacos and AdaSRacos, are presented, which utilize directional models trained on past optimization experiences to guide the search process. AdaSRacos further enhances this by adaptively selecting relevant historical experiences, ensuring that only useful information is utilized. The chapter includes empirical studies on synthetic and real-world hyper-parameter optimization tasks, demonstrating the effectiveness of the proposed methods in reducing evaluation costs and improving optimization performance. The results highlight the importance of experience adaptation in achieving efficient and effective hyper-parameter tuning.

Hyper-parameter optimization is a core task in Automatic Machine Learning (AutoML). It often follows a trial-and-error process, similar to manual tuning of hyper-parameters. The search space in hyper-parameter optimization is highly complex, being non-convex, non-differentiable, or even non-continuous. In such circumstances, derivative-free optimization (DFO) becomes a critical tool. DFO methods, such as RACOS and SRACOS introduced in Chaps. 5 and 6, also follow a trial-and-error framework. In each iteration, the optimization process samples one or several solutions, and the objective function returns their evaluation values. The optimization process then updates and samples new solutions based on the feedback.

Due to the limited optimization information (only objective function values are available), DFO methods often suffer from low efficiency, requiring a large number of samples and evaluations to achieve good optimization performance. In hyper-parameter optimization, evaluating a hyper-parameter configuration is usually expensive. For example, evaluating a learning rate setting for a deep neural network requires training the network to convergence with that learning rate, and the validation accuracy serves as the evaluation value. Therefore, accelerating DFO becomes crucial when solving hyper-parameter tuning tasks. This chapter presents an experienced

Y. Yu et al., *Derivative-Free Optimization*, Machine Learning: Foundations, Methodologies, and Applications, https://doi.org/10.1007/978-981-96-5929-6_11

optimization approach [2] that utilizes historical optimization experience to improve the optimization process for new tasks. In other words, the aim of experienced optimization is to use fewer samples to achieve better performance.

11.1 Experienced Optimization for Hyper-Parameter Optimization

Let $\mathcal{A}$ denote a learning algorithm and $\boldsymbol{\delta} \in \boldsymbol{\Delta}$ denote a hyper-parameter configuration, where $\boldsymbol{\Delta}$ is the hyper-parameter space. In machine learning, k-fold cross-validation is a common process for evaluating the quality of a learning model. The same process can be used to evaluate a hyper-parameter configuration:

$$f(\boldsymbol{\delta}) = \frac{1}{k}\sum_{i=1}^{k} \mathcal{L}(\mathcal{A}_{\boldsymbol{\delta}}, \mathcal{D}^i_{\text{train}}, \mathcal{D}^i_{\text{valid}}), \tag{11.1}$$

where $\mathcal{L}(\cdot)$ is a loss function and $\mathcal{D}^i_{\text{train}}$ and $\mathcal{D}^i_{\text{valid}}$ are the training and validation datasets, respectively, in the i-th fold. Based on the loss function definition, hyper-parameter optimization can be defined as $\boldsymbol{\delta}^* = \arg\min_{\boldsymbol{\delta}\in\boldsymbol{\Delta}} f(\boldsymbol{\delta})$.

The evaluation process of $\boldsymbol{\delta}$ often has a high time cost because it involves model training. When solving hyper-parameter optimization using DFO methods, the optimization efficiency is challenged. We note that hyper-parameter optimization tasks are related to each other. For example, when tuning hyper-parameters for a learning algorithm on different datasets, there are different optimization tasks. Due to the same learning algorithm, the hyper-parameter space of these tasks can be aligned, and tasks are similar to each other.

The experienced optimization approach presented in this chapter aims to utilize the relationships among similar hyper-parameter optimization tasks to accelerate convergence. We consider a set of optimization tasks denoted by $F = \{f\}$, where $f \sim \mathcal{F}$ and $\mathcal{F}$ is an underlying task distribution. We split F into two parts: $F = F_e \cup F_t$ and $F_e \cap F_t = \emptyset$. F_e is the set of experienced tasks that have been optimized, and F_t is the set of target tasks that have not been optimized yet. Under this problem setting, we present the ExpSRacos and AdaSRacos algorithms [2] based on SRacos (Chap. 6).

11.2 The ExpSRacos and AdaSRacos Algorithms

We observe that the search direction can be aligned and generalized across different hyper-parameter optimization tasks. In gradient-based optimization, the gradient indicates the search direction, which is key to efficient search. In DFO, we aim to learn the search direction from the experience of historical optimization processes

Algorithm 11.1 Framework of Experienced Optimization by Directional Model

Require:
F_e, F_t: Experienced and target problem sets
$\mathcal{O}$: The optimization approach
`Log&Assign`: Log and assign experience dataset
`Train`: Train directional model

Ensure:
1: $\mathcal{D}_{F_e} = \emptyset$
2: **for** $f \in F_e$ **do**
3: $\quad \mathcal{D}_{F_e}^{f} = \texttt{Log\&Assign}(\mathcal{O}, f)$
4: $\quad \mathcal{D}_{F_e} = \mathcal{D}_{F_e} \cup \mathcal{D}_{F_e}^{f}$
5: **end for**
6: $\Phi = \texttt{Train}(\mathcal{D}_{F_e})$
7: **for** $f \in F_t$ **do**
8: $\quad \boldsymbol{x}_f^* = \mathcal{O}(f, \Phi)$
9: **end for**

and use it to accelerate the search process in unseen tasks. Thus, we present the experienced optimization framework using a directional model, as presented in Algorithm 11.1.

The experienced optimization framework consists of three main steps:

- Organizing the experience dataset $\mathcal{D}_{F_e}$ from historical optimization processes (lines 1–4). DFO methods often store some historical samples during optimization. The instances in $\mathcal{D}_{F_e}$ can be extracted from snippets of the stored samples. For each instance, we extract features and assign a label indicating the direction to a later-found better solution. In line 3, the `Log&Assign` sub-process collects the labeled instances from optimization processes.
- Learning the directional model Φ on $\mathcal{D}_{F_e}$ (line 6). With the labeled experience dataset, training Φ is a supervised learning problem. Note that Φ can be trained using any state-of-the-art learning algorithm.
- Utilizing Φ to predict the direction of the next sample during optimization in new problems (lines 7–9). Directional models can be embedded in the optimization method by adding a pre-sampling step that generates a set of candidate samples. Among the candidate samples, the one closest to the direction predicted by Φ is selected as the next sample.

We first present the implementation of this framework with the SRACOS algorithm, called EXPSRACOS, and then present an improved version, ADASRACOS.

11.2.1 EXPSRACOS

In SRACOS (Algorithm 6.1 in Chap. 6), the foundation of optimization consists of two solution sets: B^+ and B^-. B^+ consists of the top-k best solutions, called positive solutions, while B^- consists of the remaining solutions, called negative solutions. ExpSRacos follows the three steps of the experienced optimization framework.

Collecting the experience dataset. At this step, we extract the experience dataset from optimization processes on previous tasks. We note that the new solution is sampled based on $(\boldsymbol{x}_t^+, B_t^-)$ in the t-th iteration, where $\boldsymbol{x}_t^+ \in B_t^+$. B_t^- stores the negative solutions. Therefore, we can organize the experience dataset using $(\boldsymbol{x}_t^+, B_t^-)$ as a context matrix:

$$\boldsymbol{\kappa}_t = \begin{bmatrix} \boldsymbol{x}_{t,1}^- - \boldsymbol{x}_t^+ \\ \boldsymbol{x}_{t,2}^- - \boldsymbol{x}_t^+ \\ \vdots \\ \boldsymbol{x}_{t,m}^- - \boldsymbol{x}_t^+ \end{bmatrix}, \text{ where } \boldsymbol{x}_{t,i}^- \in B_t^-, i = 1, 2, \ldots, m. \tag{11.2}$$

$\boldsymbol{\kappa}_t$ is an $m \times n$ matrix, where $m = |B_t^-|$ and n is the dimensionality of the search space. Each row of $\boldsymbol{\kappa}_t$ is a solution from B_t^- centralized by $\boldsymbol{x}_t^+$. The centralization aligns the search behavior at different times and in different optimization tasks, making it the key to the generalization of the directional model on unseen tasks.

Let $\boldsymbol{x}_t'$ denote the new solution sampled using the context matrix $\boldsymbol{\kappa}_t$. We combine them to create an instance of the experience dataset: $[\boldsymbol{\kappa}_t; \boldsymbol{x}_t']$. We assign a label to this instance according to the evaluation value of $\boldsymbol{x}_t'$. If the new solution improves the optimization performance so far, the label is positive; otherwise, it is negative:

$$\ell_t([\boldsymbol{\kappa}_t; \boldsymbol{x}_t']) = \begin{cases} 1, & f(\boldsymbol{x}_t') < f(\tilde{\boldsymbol{x}}_t) \\ 0, & f(\boldsymbol{x}_t') \geq f(\tilde{\boldsymbol{x}}_t) \end{cases}, \tag{11.3}$$

where $\tilde{\boldsymbol{x}}_t$ is the best-so-far solution. At each iteration, we obtain an experience instance. By combining them into a dataset, the experience dataset is $\mathcal{D}_{F_e} = \{([\boldsymbol{\kappa}_1; \boldsymbol{x}_1'], \ell_1), ([\boldsymbol{\kappa}_2; \boldsymbol{x}_2'], \ell_2), \ldots\}$.

Training the directional model. $\mathcal{D}_{F_e}$ is a dataset with binary labels. Thus, any classification algorithm can be applied to train on it. We note that an instance in $\mathcal{D}_{F_e}$ consists of two parts: $\boldsymbol{\kappa}$ and $\boldsymbol{x}$. $\boldsymbol{\kappa}$ is a matrix, and $\boldsymbol{x}$ is a vector. Therefore, we should reorganize the instance from $[\boldsymbol{\kappa}; \boldsymbol{x}]$ by reshaping $\boldsymbol{\kappa}$ into a vector and combining it with $\boldsymbol{x}$. In our work, we apply a simple multilayer perceptron (MLP) classifier as the directional model, denoted as Φ. The last layer of Φ maps the output to the range $[0, 1]$. The output is a score that reflects the quality of the new solution.

Combining ExpSRacos. We utilize Φ within the framework of SRacos, resulting in the ExpSRacos algorithm. Before evaluating a solution, a pre-sampling step is added to generate a set of solutions. These solutions are then filtered by the directional model. Algorithm 11.2 presents the pseudo-code of ExpSRacos. Line 1 is the initialization step. Lines 4–8 constitute the pre-sampling process. The directional model Φ predicts the quality of each pre-sampled solution (line 6). Only the solution with the highest predicted value is evaluated by the real objective function (lines 9–10) and used to update (B^+, B^-).

Algorithm 11.2 Experienced SRacos (ExpSRacos)

Require:
f: Objective function to be minimized
P: The number of pre-samples
r: The number of samples in initialization
N: The evaluation budget
Φ: Directional model
`Initialize`: Initialization steps
`Sample`: Get a new sample by SRacos
Ensure:
$(B^+, B^-, (\tilde{x}, \tilde{y})) = \texttt{Initialize}(\mathcal{U}_{\mathcal{X}})$
for $t = r + 1$ to N **do**
  $\mathcal{P} = \emptyset$
  for $i = 1$ to P **do**
    $(\boldsymbol{\kappa}, \boldsymbol{x}) = \texttt{Sample}(B^+, B^-, \lambda, \mathcal{C})$
    $p = \Phi(\boldsymbol{\kappa}, \boldsymbol{x})$
    $\mathcal{P} = \mathcal{P} \cup \{((\boldsymbol{\kappa}, \boldsymbol{x}), p)\}$
  end for
  $(\hat{\boldsymbol{\kappa}}, \hat{\boldsymbol{x}}), \hat{p} = \arg\max_{((\boldsymbol{\kappa},\boldsymbol{x}),p)\in\mathcal{P}} p$
  $\hat{y} = f(\hat{\boldsymbol{x}})$
  $(B^+, B^-) = \texttt{Update}((\hat{\boldsymbol{x}}, \hat{y}), B^+, B^-)$
  $(\tilde{\boldsymbol{x}}, \tilde{y}) = \arg\min_{(\boldsymbol{x},y)\in B^+\cup\{(\tilde{\boldsymbol{x}},\tilde{y})\}} y$
end for
return $(\tilde{x}, \tilde{y})$

We discuss why experienced optimization works under the following two assumptions:

- We assume that optimization tasks $f \in F_e$ and F_t share the same search space $\mathcal{X}$.
- For any two instances $([\boldsymbol{\kappa}_{\mathbf{a}}; \boldsymbol{x}_{\mathbf{a}}], \ell_{\mathbf{a}})$ and $([\boldsymbol{\kappa}_{\mathbf{b}}; \boldsymbol{x}_{\mathbf{b}}], \ell_{\mathbf{b}})$ in $\mathcal{D}_{F_e}$, we assume $\ell_{\mathbf{a}} = \ell_{\mathbf{b}}$ if $[\boldsymbol{\kappa}_{\mathbf{a}}; \boldsymbol{x}_{\mathbf{a}}] = [\boldsymbol{\kappa}_{\mathbf{b}}; \boldsymbol{x}_{\mathbf{b}}]$.

The centralization process of $\boldsymbol{\kappa}$ ensures that the second assumption is met in the majority of cases. The directional model Φ learned from historical optimization processes predicts whether a new solution $\boldsymbol{x}$ is good. Based on these assumptions, Φ can be reused to predict the quality of new solutions on unseen tasks. In ExpSRacos, the solution with the highest predicted value from Φ is evaluated by the real objective function. Compared to SRacos, which wastes many samples on exploration, ExpSRacos avoids evaluating many inferior solutions.

11.2.2 AdaSRacos

Experienced optimization works when the experience extracted from historical tasks can provide the right directions for unseen tasks. In real cases, we cannot guarantee that the experience from historical tasks will positively affect the optimization process on new tasks. In Fig. 11.1, the target task is the Sphere function with the optimal point

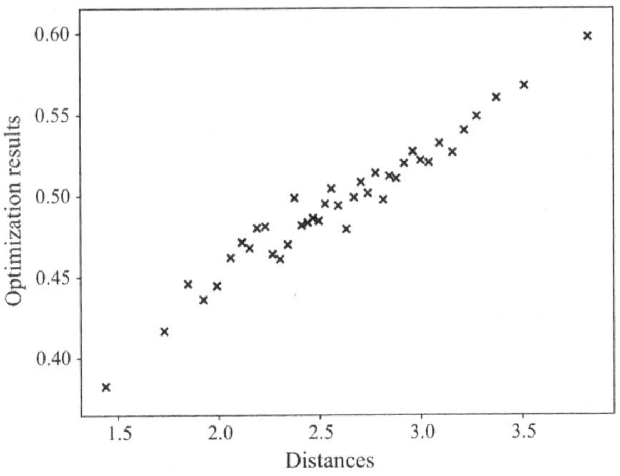

Fig. 11.1 Illustration of the relevance between experience and target tasks [2]. The task is minimizing the Sphere function. The X-axis is the Euclidean distance between the optimal points of the source and target tasks. A smaller distance indicates stronger relevance between the two tasks. The Y-axis is the performance of experienced optimization with an evaluation budget of only 50

$\{0.1\}^n$ $(n = 10)$, and we randomly shift the optimal points of the Sphere function to construct the experienced task distribution. As shown in the figure, the optimization performance of experienced optimization is strongly related to the relevance between tasks. Therefore, we need to address the problem of selecting positive experience, i.e., the experience task most relevant to the target task. This leads to the adaptive experienced optimization algorithm, AdaSRacos.

In the experienced optimization setting, $F_e = \{f_1, f_2, \ldots, f_{M_e}\}$ denotes the set of M_e experience tasks. Correspondingly, we obtain a set of directional models denoted by $\boldsymbol{\Phi} = \{\Phi_1, \Phi_2, \ldots, \Phi_{M_e}\}$. Φ_i is a directional model trained on the experience dataset obtained by optimizing f_i, where $i \in \{1, 2, \ldots, M_e\}$. We consider a weighted ensemble approach to combine all directional models:

$$\bar{\Phi}([\boldsymbol{\kappa}; \boldsymbol{x}]) = \sum_{i=1}^{M_e} w_i \Phi_i([\boldsymbol{\kappa}; \boldsymbol{x}]), \tag{11.4}$$

where $\boldsymbol{w} = \{w_1, w_2, \ldots, w_{M_e}\}$ are the weights of the directional models. The weight intuitively indicates the relevance between the experience and target tasks. We want the relevant directional model to have a larger weight. We note that the ground-truth label of a piece of experience data $([\boldsymbol{\kappa}; \boldsymbol{x}], \ell)$ can be obtained during optimization because $\boldsymbol{x}$ will be evaluated by the objective function. In the optimization process, the labeled experience data $([\boldsymbol{\kappa}; \boldsymbol{x}], \ell)$ arrives sequentially. Furthermore, the prediction of each directional model is a real number in $[0, 1]$. We define a squared loss to measure the prediction quality of the directional model: $(\Phi_i([\boldsymbol{\kappa}; \boldsymbol{x}]) - \ell)^2$. The weights of all directional models can be adapted according to the loss:

Algorithm 11.3 DFO with Adaptive Experience (AdaSRacos)

Require: (extra input compared to Algorithm 11.2)
  $\mathbf{\Phi} = \{\Phi_1, \Phi_2, \ldots, \Phi_{M_e}\}$: Basic directional model set
  `Normalize`: A normalization procedure
Ensure:
1: $B, \tilde{\boldsymbol{x}} = \texttt{Initialize}(\mathcal{U}_\mathcal{X})$
2: $\boldsymbol{w} = \{w_1, w_2, \ldots, w_{M_e}\} = \{\frac{1}{M_e}\}^{M_e}$
3: **for** $t = 1$ to N **do**
4: $\quad \mathcal{P} = \emptyset$
5: $\quad$ **for** $p = 1$ to P **do**
6: $\quad\quad [\boldsymbol{\kappa}_p; \boldsymbol{x}_p] = \texttt{Sample}(B)$
7: $\quad\quad \mathcal{P} = \mathcal{P} \cup \{[\boldsymbol{\kappa}_p; \boldsymbol{x}_p]\}$
8: $\quad$ **end for**
9: $\quad [\boldsymbol{\kappa}'; \boldsymbol{x}'] = \arg\max_{[\boldsymbol{\kappa};\boldsymbol{x}]\in\mathcal{P}} \sum_{i=1}^{M_e} w_i \Phi_i([\boldsymbol{\kappa}; \boldsymbol{x}])$
10: $\quad \ell' = \begin{cases} 1, & f(\boldsymbol{x}') < f(\tilde{\boldsymbol{x}}) \\ 0, & f(\boldsymbol{x}') \geq f(\tilde{\boldsymbol{x}}) \end{cases}$
11: $\quad$ **for** $i = 1$ to M_e **do**
12: $\quad\quad w_i = \exp(-\alpha(\Phi_i([\boldsymbol{\kappa}'; \boldsymbol{x}']) - \ell')^2) w_i$
13: $\quad$ **end for**
14: $\quad \boldsymbol{w} = \texttt{Normalize}(\boldsymbol{w})$
15: $\quad B = \texttt{Update}(B, \boldsymbol{x}', f(\boldsymbol{x}'))$
16: $\quad$ **if** $f(\boldsymbol{x}') < f(\tilde{\boldsymbol{x}})$ **then**
17: $\quad\quad \tilde{\boldsymbol{x}} = \boldsymbol{x}'$
18: $\quad$ **end if**
19: **end for**
20: **return** $\tilde{\boldsymbol{x}}$

$$w_i = \exp(-\alpha(\Phi_i([\boldsymbol{\kappa}; \boldsymbol{x}]) - \ell)^2) w_i, \tag{11.5}$$

where α is a scale hyper-parameter. Based on ExpSRacos (Algorithm 11.2), we apply the above weight adaptation mechanism to construct the adaptive ExpSRacos algorithm, AdaSRacos (Algorithm 11.3).

Algorithm 11.3 still follows the pre-sampling mechanism to utilize the directional model. The algorithm starts with optimization initialization. We set the same weight $\frac{1}{M_e}$ for all basic directional models (line 2). Lines 5–8 constitute the pre-sampling phase. We utilize the weighted ensemble directional model to predict the quality score for each temporary solution (line 9). The solution with the highest predicted value is evaluated by the real objective function (lines 10 and 15). We adapt the weights for all basic directional models during lines 11–14. First, we obtain the ground-truth label for the experience data $[\boldsymbol{\kappa}'; \boldsymbol{x}']$ (line 10). Then, we adjust the weight for each directional model based on the prediction loss (lines 11–13). Finally, we apply a normalization procedure to ensure that $\sum_{i=1}^{M_e} w_i = 1$. With the selected solution and its evaluation value, we update the optimization procedure (line 15) and the best-so-far solution (lines 16–18). When the evaluation budget is exhausted, the best-so-far solution is returned (line 20).

We discuss the experience adaptation mechanism based on experienced DFO. With the weighted ensemble, all basic directional models obtained from different experience tasks are integrated into a single directional model. When optimizing on target tasks, we first employ the ensemble directional model to select solutions worth evaluating. We then test all basic directional models on the labeled experience data. According to the test results, the relevant directional models are selected by adapting the weights. If a basic directional model gives a correct prediction, it will obtain a small squared loss, and its corresponding weight will receive a small discount. However, when a basic directional model gives a wrong prediction, its weight will be heavily discounted. After the normalization step, the weights of basic directional models that make correct predictions will increase relatively, while the weights of those that make wrong predictions will decrease relatively. In this way, relevant directional models that make fewer mistakes on the target task can be adaptively selected with large weights, and irrelevant directional models that make more mistakes can be adaptively omitted with small weights.

11.3 Empirical Study

We compare ExpSRacos [2] and AdaSRacos [3] to several state-of-the-art DFO methods, including SRacos [1], SMAC [4], and Bayes [6]. We apply an MLP classifier as the directional model for ExpSRacos and AdaSRacos, with the network structure depending on the tasks.

11.3.1 Synthetic Tasks

We select the Sphere and Rosenbrock functions as the basic synthetic functions. The Sphere function is convex:

$$f(\boldsymbol{x}) = \sum_{i=1}^{n} (x_i - x_i^*)^2. \tag{11.6}$$

The Rosenbrock function is non-convex:

$$f(\boldsymbol{x}) = \sum_{i=1}^{n-1} [100(x_{i+1} - x_{i+1}^* - (x_i - x_i^*)^2)^2 + (1 - x_i + x_i^*)^2], \tag{11.7}$$

where $\boldsymbol{x} = \{x_1, x_2, \ldots, x_n\}$ is a solution, n is the dimensionality, and $\boldsymbol{x}^* = \{x_1^*, x_2^*, \ldots, x_n^*\}$ is the optimal point. The task is to minimize the synthetic function value within a constrained region. The relevance between two tasks can be easily measured by the distance between their optimal points. With this relevance, we can investigate the effectiveness of the presented experienced optimization and the weight adaptation mechanism.

Table 11.1 Average performance on synthetic target tasks [2], i.e., Sphere and Rosenbrock functions with the optimal points $\boldsymbol{x}^* = \{0.10\}^{10}$, $\{0.25\}^{10}$, and $\{0.40\}^{10}$. The bold values represent the best performance

Function	$\boldsymbol{x}^*$	AdaSRacos		ExpSRacos	
		Sphere-Set	Mixed-Set	Sphere-Set	Mixed-Set
Sphere	$\{0.10\}^{10}$	**0.0694** ± 0.02	0.0747 ± 0.02	0.1132 ± 0.05	0.1165 ± 0.07
Sphere	$\{0.25\}^{10}$	**0.0775** ± 0.03	0.1165 ± 0.07	0.1091 ± 0.05	0.1250 ± 0.08
Sphere	$\{0.40\}^{10}$	**0.0909** ± 0.04	0.1528 ± 0.22	0.1978 ± 0.05	0.2938 ± 0.19
Rosenbrock	$\{0.10\}^{10}$	12.394 ± 2.27	**11.010** ± 0.69	13.351 ± 3.39	14.109 ± 4.62
Rosenbrock	$\{0.25\}^{10}$	25.549 ± 9.54	**15.814** ± 6.62	26.008 ± 8.93	17.771 ± 3.16
Rosenbrock	$\{0.40\}^{10}$	57.388 ± 20.4	**45.408** ± 34.8	93.370 ± 40.7	54.763 ± 22.6

Function	SRacos	SMAC	Bayes
Sphere $\{0.10\}^{10}$	0.7941 ± 0.29	0.0700 ± 0.01	0.4894 ± 0.05
Sphere $\{0.25\}^{10}$	0.8046 ± 0.39	0.2749 ± 0.11	0.4500 ± 0.11
Sphere $\{0.40\}^{10}$	0.8306 ± 0.36	0.6778 ± 0.25	0.3444 ± 0.09
Rosenbrock $\{0.10\}^{10}$	26.903 ± 5.18	17.176 ± 1.15	45.523 ± 17.1
Rosenbrock $\{0.25\}^{10}$	33.065 ± 29.1	43.701 ± 8.37	45.733 ± 13.9
Rosenbrock $\{0.40\}^{10}$	61.955 ± 24.2	99.798 ± 43.6	48.504 ± 12.9

Task setting. We construct the task distribution $\mathcal{F}$ by randomly shifting the optimal point for the Sphere and Rosenbrock functions. The shifting region is $[-0.5, 0.5]$. We have two kinds of experience task sets: Sphere set and Mixed set. The Sphere set contains 20 Sphere functions with 20 different optimal points. The Mixed set contains 10 Sphere functions with 10 different optimal points and 10 Rosenbrock functions with 10 different optimal points. For the target tasks, we select three optimal points: $\boldsymbol{x}^* = \{0.1\}^n, \{0.25\}^n, \{0.4\}^n$. By combining them with the Sphere and Rosenbrock functions, we have a total of six target tasks. The search space is $[-1, 1]^n$, where $n = 10$. The number of evaluations, i.e., the budget, is set to 50 for all compared methods.

Directional model training. We optimize the tasks in the Sphere set and Mixed set separately to collect experience datasets. The budget is 500, and the optimization process is repeated 10 times. The directional model is an MLP classifier. Thus, we obtain 20 directional models from the Sphere set and 20 directional models from the Mixed set. For ExpSRacos, we construct a directional model by an average weighted ensemble. For AdaSRacos, we use the weighted ensemble approach, but the weights are adapted using the presented adaptation mechanism.

Empirical analysis. We independently repeat each experiment on the target tasks 10 times and report the average performance in Table 11.1. The table includes two subtables. The first subtable shows the performance of experienced optimization, i.e., ExpSRacos and AdaSRacos. The second subtable shows the performance of classical DFO methods, i.e., SRacos, SMAC, and Bayes. AdaSRacos achieves the best performance on all six target tasks, indicating that experienced optimization with the adaptation mechanism is powerful on unseen tasks. Comparing experienced optimization (first subtable) to classical DFO (second subtable), experienced optimization generally outperforms classical DFO. We further verify how the weights of AdaSRacos change during optimization by showing the weight changes in Fig. 11.2. In each figure, the X-axis from left to right indicates decreasing relevance. The weight bars on the left become higher, indicating that relevant directional models are effectively selected. The weight bars on the right become lower, indicating that irrelevant directional models are effectively omitted.

11.3.2 Hyper-Parameter Optimization Tasks

The classifier selected for hyper-parameter optimization tasks is LightGBM [5]. We select a total of 11 hyper-parameters, including boosting type, learning rate, number of estimators, number of leaves, etc. We use 40 datasets and F1 score as the evaluation

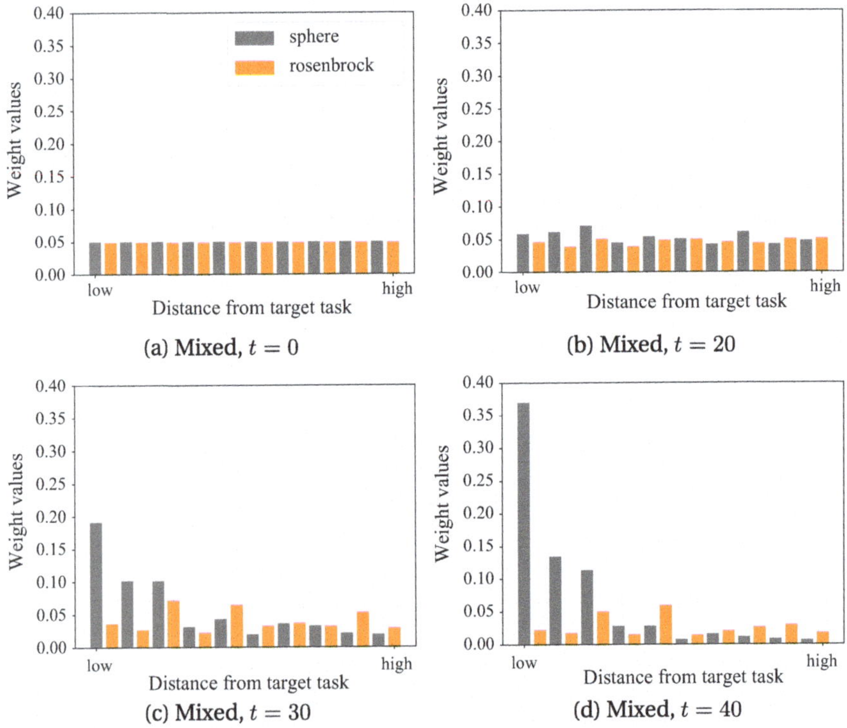

Fig. 11.2 Illustration of AdaSRacos weight changes on the target task [2], i.e., the 10-dimensional Sphere function with the optimal point $\boldsymbol{x}^* = \{0.1\}^{10}$. Weights are from AdaSRacos with Mixed set experience at optimization steps $t = 0, 20, 30,$ and 40. In each figure, the X-axis from left to right indicates increasing distances between the optimal points of the experience and target tasks

criterion. The goal of hyper-parameter optimization is to maximize the F1 score by tuning the hyper-parameters of LightGBM on the 40 datasets.

Task setting. The 40 datasets are split into two parts: 30 experience task datasets and 10 target task datasets. To train the directional model, we employ SRACOS to optimize the hyper-parameters of LightGBM with 300 evaluations. Each optimization process is repeated 10 times. The running log constructs the experience dataset for training the directional model. The basic directional model is an MLP classifier. We set the evaluation budget to 30 for all experiments and independently run each experiment 5 times. We report the average performance.

Table 11.2 shows the hyper-parameter optimization performance of the compared methods on the 40 datasets. We compare the hyper-parameter optimization performance to the baseline, i.e., LightGBM with default hyper-parameters. Hyper-parameter optimization outperforms the baseline on 38 datasets, indicating that hyper-parameter optimization is necessary for machine learning applications.

On experience datasets, we compare EXPSRACOS to basic DFO methods (SRACOS, SMAC, and Bayes). EXPSRACOS is even worse than SRACOS and SMAC. This phenomenon is caused by irrelevant directional models that have negative impacts on EXPSRACOS. ADASRACOS outperforms the other compared methods on 29/30 datasets and obtains an average rank of 1.13. This indicates that selecting relevant directional models is necessary for improving hyper-parameter optimization performance, and the presented experience adaptation mechanism can effectively eliminate the negative impact of irrelevant basic directional models and correctly select the relevant directional models.

On target datasets, the optimization results show that ADASRACOS can effectively transfer optimization experience to unseen tasks (ADASRACOS outperforms the other compared methods on all 10 datasets). EXPSRACOS beats SRACOS on 9/10 datasets, showing that experienced optimization is helpful for improving optimization performance on unseen tasks. However, compared with ADASRACOS, ADASRACOS significantly improves the F1 score on all 10 datasets with only an evaluation budget of 30. This indicates that experience adaptation can significantly improve the efficiency of hyper-parameter optimization.

11.4 Summary

Hyper-parameter optimization plays an important role in AutoML. A classical solver employs DFO to discover the hyper-parameter configuration with the best performance. Due to the high evaluation cost, previous hyper-parameter optimization methods suffer from low efficiency, i.e., they require a long time to find a sufficiently good hyper-parameter configuration. To address this issue, this chapter presents the experienced optimization algorithm EXPSRACOS, which utilizes the experience of historical optimization processes to accelerate new optimization on target tasks. However, irrelevant experience can have a negative impact on experienced optimization. This

Table 11.2 Average optimization F1 score of hyper-parameter optimization on 40 datasets [2]. B.L. means baseline that is the F1 score of LightGBM with default hyper-parameters. The numbers with • and ◦ are the first and second rank performances. We also analyze the number of first/second/third ranks and average rank (Avg. Rank)

	Dataset	Optimization F1 score on training dataset					B.L.
		AdaSRA.	ExpSRA.	SRacos	SMAC	Bayes	
Experienced datasets	Australian	.9026•	.8817	.8871◦	.8871◦	.8724	.8389
	Breast	.9999•	.9999•	.9999•	.9999•	.9999•	.9402
	Electricity	.7431•	.7398◦	.7377	.7345	.7322	.5492
	Buggy.C.	.8943•	.8693	.8825	.8864◦	.8825	.8552
	CMC	.5860•	.5738	.5754◦	.5715	.5741	.4614
	Contrac.	.5785•	.5762◦	.5750	.5715	.5725	.4614
	Credit.A.	.8938•	.8921	.8927◦	.8864	.8895	.8250
	G.E.2-1000	.5772•	.5433	.5518	.5590◦	.5378	.5368
	G.E.2-200	.7534•	.7040	.7041	.7534•	.7131	.6187
	G.E.3-20	.5784•	.5623	.5657◦	.5485	.5601	.4936
	G.H.20	.7221•	.7040	.7169	.7187◦	.7076	.6747
	H.V.wo.N.	.5934•	.5875	.5786	.5642	.5888◦	.5977
	H.V.w.N.	.5931•	.5889	.5871	.5910◦	.5823	.5241
	Mfeat.K.	.9713•	.9713•	.9680	.9692	.9693	.9197
	Mfeat.M.	.7235•	.7212◦	.7161	.7140	.7212◦	.6967
	Mfeat.P.	.9722•	.9674	.9685	.9721◦	.9661	.9501
	Mfeat.Z.	.7867•	.7758	.7780◦	.7771	.7766	.7411
	Monk2	.8732•	.6981	.6548	.5774	.7413◦	.6089
	Parity5.	.4948•	.4815	.4847◦	.4820	.4837	.2291
	Pima	.7236•	.7102	.6948	.7173	.7177◦	.6590
	Tic.T.T	.9741•	.9163	.9401	.9741•	.9241	.7898
	Tokyo.1	.9248•	.9203	.9168	.9243◦	.9210	.9081
	Vehicle	.7943•	.7853	.7911◦	.7763	.7814	.7610
	Wine.Q.R.	.4218•	.3875	.3617	.4021◦	.3620	.2589
	Yeast	.4754•	.4435	.4447◦	.4388	.4388	.4716
	Airlines	.6488•	.6483◦	.6467	.6438	.6464	.5943
	Titanic	.8238•	.8221◦	.8187	.8099	.8048	.8217
	Twonorm	.9749	.9750	.9751	.9782•	.9757	.9541
	Glass	.7499•	.7088	.7125	.7071	.7497◦	.4345
	Horse.C.	.8724•	.8586	.8602	.8702◦	.8616	.7989
1st/2nd/3rd		29/0/0	2/5/8	1/8/11	4/9/2	1/6/7	–
Avg.Rank		1.13	3.43	3.13	3.27	3.47	–
Target Datasets	Messidor	.7548•	.7525◦	.7462	.7353	.7505	.6581
	Adult	.8137•	.8121	.8104	.8128	.8129◦	.7558
	Balance.S.	.5448•	.5399	.5380	.5409◦	.5398	.5294
	CNAE	.9060•	.8924	.8955◦	.8946	.8920	.8227
	Credit.G	.7190•	.7174◦	.7052	.7173	.7168	.6894
	CRX	.8864•	.8854◦	.8843	.8690	.8810	.8974
	Cylinder	.8094•	.8065◦	.7487	.7791	.7953	.7990
	Flare	.7174•	.6816	.6704	.7141◦	.6954	.4518
	Solar.F.	.6724•	.6233	.6131	.6448◦	.6195	.5758
	German	.7498•	.7457	.7331	.7463◦	.7432	.5482
1st/2nd/3rd		10/0/0	0/4/3	0/1/1	0/4/3	0/1/3	–
Avg.Rank		1.00	2.90	4.40	3.10	3.60	–

chapter further presents an experience adaptation mechanism that tests the experience on target tasks. The relevant experience that makes fewer mistakes is adaptively selected, while the irrelevant experience that makes more mistakes is omitted. We implement AdaSRacos based on ExpSRacos. Experiments on synthetic tasks verify that AdaSRacos can effectively discover the relevance among tasks. The empirical results of hyper-parameter optimization on 40 datasets show that AdaSRacos significantly improves the efficiency of hyper-parameter optimization.

References

1. Hu YQ, Qian H, Yu Y (2017) Sequential classification-based optimization for direct policy search. In: Proceedings of the 31st AAAI conference on artificial intelligence, San Francisco, CA, pp 2029–2035
2. Hu YQ, Yu Y, Zhou ZH (2018) Experienced optimization with reusable directional model for hyper-parameter search. In: Proceeding of the 27th international joint conference on artificial intelligence, pp 2276–2282
3. Hu YQ, Liu Z, Yang H, Yu Y, Liu Y (2020) Derivative-free optimization with adaptive experience for efficient hyper-parameter tuning. In: Proceeding of the 24th European conference on artificial intelligence, pp 1207–1214
4. Hutter F, Hoos HH, Leyton-Brown K (2011) Sequential model-based optimization for general algorithm configuration. LION 5:507–523
5. Ke G, Meng Q, Finley T, Wang T, Chen W, Ma W, Ye Q, Liu TY (2017) Lightgbm: a highly efficient gradient boosting decision tree. In: Advances in neural information processing systems, pp 3146–3154
6. Shahriari B, Swersky K, Wang Z, Adams RP, Freitas ND (2015) Taking the human out of the loop: a review of bayesian optimization. Proc IEEE 104(1):148–175

Chapter 12
Multi-fidelity Optimization: Acceleration in Hyper-Parameter Evaluation

Abstract This chapter addresses the challenge of expensive evaluations in hyper-parameter optimization by introducing a multi-fidelity optimization approach. Hyper-parameter optimization often involves time-consuming evaluations, especially with large datasets or complex models. The chapter proposes a method that combines low-fidelity evaluations (using subsets of data) with high-fidelity evaluations (using full datasets) to accelerate the optimization process. A key innovation is the Transfer Series Expansion (TSE) algorithm, which predicts the residual between low and high-fidelity evaluations, allowing for efficient optimization with fewer costly evaluations. The chapter presents the TseSRacos algorithm, which integrates TSE with the SRacos optimization framework. Empirical studies on LightGBM hyper-parameter optimization tasks demonstrate that the proposed method significantly reduces evaluation time while maintaining high optimization performance. The results highlight the effectiveness of multi-fidelity optimization in improving efficiency, particularly for large-scale datasets. The chapter concludes that TSE-based multi-fidelity optimization is a powerful tool for accelerating hyper-parameter tuning in machine learning.

In this chapter, we continue to focus on hyper-parameter optimization tasks. As discussed in Chap. 11, derivative-free optimization (DFO) is a popular solver for hyper-parameter optimization. Hyper-parameter optimization typically involves a large number of iterations, each of which includes a hyper-parameter evaluation process that requires training the model and validating it on the real dataset. When the dataset is large or the model is complex, the evaluation process becomes very time-consuming, leading to inefficiency in hyper-parameter optimization.

Chapter 11 addressed the inefficiency of hyper-parameter optimization by reducing the number of iterations. This chapter tackles the problem from another perspective: accelerating the hyper-parameter evaluation [3], using a multi-fidelity approach. By shortening the time cost of each iteration, the optimization process can increase the number of iterations to improve optimization performance.

Y. Yu et al., *Derivative-Free Optimization*, Machine Learning: Foundations, Methodologies, and Applications, https://doi.org/10.1007/978-981-96-5929-6_12

12.1 Multi-fidelity Optimization for Hyper-Parameter Optimization

We note that it is easy to construct multi-fidelity evaluations when optimizing hyper-parameters. For example, when evaluating hyper-parameters on a small subset of the dataset, it is considered a low-fidelity evaluation. Low-fidelity evaluations are much faster than evaluations on the full dataset, but they are less accurate. High-fidelity evaluations, which train the model on the full dataset, are highly time-consuming but provide an accurate measure of the hyper-parameters' quality. This raises the question: can we combine different fidelity evaluations to accelerate the evaluation processes while obtaining accurate evaluation values when optimizing hyper-parameters? This leads to the concept of multi-fidelity DFO [1, 8]. Multi-fidelity optimization has been extensively studied in the context of surrogate model optimization, such as Bayesian optimization [4–6]. However, few works [9] have explored its application to other optimization methods.

Let us review the notations for hyper-parameter optimization introduced in Chap. 11. $\mathcal{A}$ denotes an algorithm, and $\boldsymbol{\delta} \in \boldsymbol{\Delta}$ denotes a hyper-parameter configuration, where $\boldsymbol{\Delta}$ is the corresponding hyper-parameter space. The evaluation process of a hyper-parameter configuration is a k-fold cross-validation process:

$$f(\boldsymbol{\delta}) = \frac{1}{k}\sum_{i=1}^{k} \mathcal{L}(\mathcal{A}_{\boldsymbol{\delta}}, \mathcal{D}^i_{\text{train}}, \mathcal{D}^i_{\text{valid}}), \tag{12.1}$$

where $\mathcal{L}(\cdot)$ is a loss function, and $\mathcal{D}^i_{\text{train}}$ and $\mathcal{D}^i_{\text{valid}}$ are the training and validation datasets in the ith fold. The goal of hyper-parameter optimization is to minimize the objective function $f(\cdot)$.

In the optimization setting, we still use $\boldsymbol{x} \in \mathcal{X}$ to denote a solution, where $\mathcal{X}$ is the search space. $\boldsymbol{x}$ corresponds to $\boldsymbol{\delta}$ when optimizing hyper-parameters. In multi-fidelity optimization, a solution $\boldsymbol{x}$ can be evaluated at several different levels. We consider the simplest situation, where there are only two evaluation functions: $f_H : \mathcal{X} \to \mathbb{R}$ denotes the high-fidelity evaluation function, which outputs an accurate evaluation value for solutions but has a high time cost, and $f_L : \mathcal{X} \to \mathbb{R}$ denotes the low-fidelity evaluation function, which quickly outputs an evaluation value but with a strong bias. In hyper-parameter optimization, the high-fidelity function is

$$f_H(\boldsymbol{\delta}) = \mathcal{L}(\boldsymbol{\delta}, \mathcal{D}_{\text{train}}, \mathcal{D}_{\text{valid}}). \tag{12.2}$$

The full dataset is used to evaluate the quality of a hyper-parameter configuration $\boldsymbol{\delta}$. When training the algorithm on a small subset of the dataset, it constructs a low-fidelity function:

$$f_L(\boldsymbol{\delta}) = \mathcal{L}(\boldsymbol{\delta}, \mathcal{D}^{r_L}_{\text{sub}}, \mathcal{D}_{\text{valid}}), \tag{12.3}$$

where $0 < r_L \ll 1$ is the subsample ratio that indicates the size of the data subset, and $\mathcal{D}_{\text{sub}}^{r_L} \subset \mathcal{D}_{\text{train}}$ is randomly selected.

In this chapter, we present a method that learns a model to predict the residual between f_H and f_L with few observations. During optimization, we only use f_L as the evaluation function, which is corrected by this predicted model. Due to the cheap evaluation function, the optimization process is accelerated. Based on this idea, we present the *Transfer Series Expansion* (TSE) optimization method and implement it following the SRacos algorithm (Chap. 6), naming it TseSRacos [3].

12.2 The TseSRacos Algorithm

The bias of f_L is the biggest obstacle preventing us from directly using the low-fidelity evaluation function. We define a simple regret measuring the residual between f_H and f_L as follows:

$$R(\boldsymbol{x}) = f_H(\boldsymbol{x}) - f_L(\boldsymbol{x}), \tag{12.4}$$

where $\boldsymbol{x}$ is a solution sampled from the search space. In hyper-parameter optimization tasks, $\boldsymbol{x}$ is equivalent to $\boldsymbol{\delta}$. Based on this regret definition, once R is available, we can use $R + f_L$ to substitute for f_H. In real optimization, we have some opportunities to evaluate $\boldsymbol{x}$ by f_H. Thus, we can create a dataset $\mathcal{D} = \{(\boldsymbol{x}_1, y_1), (\boldsymbol{x}_2, y_2), \ldots\}$, where $y_i = R(\boldsymbol{x}_i)$. The regret function can be easily learned by a supervised regression learner Ψ from $\mathcal{D}$. Because $f_H(\boldsymbol{x})$ has a high time cost, the size of $\mathcal{D}$ is limited. Therefore, the challenge of learning Ψ is to train an accurate regret predictor from a very small supervised dataset.

12.2.1 *Multi-fidelity Optimization Framework*

The regret predictor-based multi-fidelity optimization is a general framework that can be implemented with any DFO method. We focus on minimization problems. Let $\texttt{Sample}_{\mathcal{O}}$ denote the step of generating a new sample, which is a key step for a DFO method, where $\mathcal{O}$ is a DFO method. Different methods have different `Sample` steps. In multi-fidelity optimization, the low-fidelity evaluation f_L is introduced to decrease the total evaluation cost. The optimization process more frequently evaluates solutions using f_L. The regret predictor Ψ is a core component of this framework. The dataset for training Ψ requires some samples evaluated by the high-fidelity evaluation function f_H. Thus, we also need a sub-procedure `Find` to select a sample to be evaluated by f_H from the sample set. The presented framework learns a predictor Ψ to estimate the residual between high and low-fidelity evaluations. Then, optimization using the corrected evaluations $(f_L + \Psi)$ can find a good sample that still performs well under high-fidelity evaluation.

Algorithm 12.1 Multi-fidelity Optimization Framework

Input:

$\mathcal{X}$: The optimization space
f_L, f_H: Low and high-fidelity evaluation functions
T_H: The budget of high-fidelity evaluations
T_L: The low-fidelity evaluation times between high-fidelity evaluations
`Initialization`: Initialization step
$\texttt{Sample}_{\mathcal{O}}$: The sample step in optimization method $\mathcal{O}$
`Find`: Select a sample to be evaluated by f_H
`Train`: Train the regret predictor

Procedure:

1: $\mathcal{X}_H, \boldsymbol{y} = \emptyset, \emptyset$
2: $\Psi = \texttt{Initialize}, \forall \boldsymbol{x} \in \mathcal{X} : \Psi(\boldsymbol{x}) = 0$
3: $\mathcal{X}_L = \texttt{Initialization}(\mathcal{X})$
4: **for** $t_H = 1$ to T_H **do**
5:     **for** $t_L = 1$ to T_L **do**
6:         $\boldsymbol{x} = \texttt{Sample}_{\mathcal{O}}(\mathcal{X}, f_L + \Psi)$
7:         $\mathcal{X}_L = \mathcal{X}_L \cup \{\boldsymbol{x}\}$
8:     **end for**
9:     $\boldsymbol{x}' = \texttt{Find}(\mathcal{X}_L)$
10:     $\gamma' = f_H(\boldsymbol{x}')$
11:     $\mathcal{X}_H, \boldsymbol{y} = \mathcal{X}_H \cup \{\boldsymbol{x}'\}, \boldsymbol{y} \cup \{\gamma' - f_L(\boldsymbol{x}')\}$
12:     $\Psi = \texttt{Train}(\mathcal{X}_H, \boldsymbol{y})$
13: **end for**
14: **return** $\arg\min_{\boldsymbol{x} \in \mathcal{X}_H} f_H(\boldsymbol{x})$

Algorithm 12.1 presents the general multi-fidelity optimization framework. $\mathcal{X}_H$ is a sample set that stores $\boldsymbol{x}$ evaluated by f_H. $\boldsymbol{y}$ is a set of regression targets corresponding to $\mathcal{X}_H$. $\mathcal{X}_H$ and $\boldsymbol{y}$ are initially empty (line 1). In the initialization step (line 2), Ψ can only output 0 because there is no learning information. $\mathcal{X}_L$ is a set to store all samples generated by the method. The initialization step (line 3) samples from $\mathcal{X}$ uniformly. In each iteration, first, the corrected low-fidelity evaluation $(f_L + \Psi)$ is considered as the objective function, and T_L samples are generated in this loop (lines 5–8). Then, a sub-process `Find` selects a sample to be evaluated by the high-fidelity function f_H (lines 9-10). With the high-fidelity evaluated sample, this framework constructs the regression dataset (line 11) to re-train the predictor Ψ (line 12). As t_H increases, Ψ approaches the real simple regret function. Optimization using the corrected low-fidelity evaluation is similar to optimization using the high-fidelity function. Finally, the algorithm returns the best-so-far sample (line 14).

12.2.2 Transfer Series Expansion (TSE)

A major challenge of Algorithm 12.1 is that Ψ is not accurate enough because the training dataset is small. To address this, we present the Transfer Series Expansion (TSE) algorithm. We assume there is a series of pre-trained base predictors

Algorithm 12.2 Multi-fidelity Optimization with TSE

Input: (extra input compared to Algorithm 12.1)
  $\boldsymbol{\psi} = \{\psi_1, \psi_2, \ldots, \psi_k\}$: The base predictor set
Procedure:
$\mathcal{X}_H, \mathbf{Z}, \boldsymbol{y} = \emptyset, \emptyset, \emptyset$
$\Psi_{\boldsymbol{w}}^{\boldsymbol{\psi}} = \Psi_{\mathbf{0}}^{\boldsymbol{\psi}}$
$\mathcal{X}_L = \texttt{Initialization}(\mathcal{X})$
$(\tilde{\boldsymbol{x}}, \tilde{\gamma}) = (\mathbf{0}, +\infty)$
for $t_H = 1$ to T_H **do**
  for $t_L = 1$ to T_L **do**
    $\boldsymbol{x} = \texttt{Sample}_{\mathcal{O}}(\mathcal{X}_L, f_L + \Psi_{\boldsymbol{w}}^{\boldsymbol{\psi}})$
    $\mathcal{X}_L = \mathcal{X}_L \cup \{\boldsymbol{x}\}$
  end for
  $\boldsymbol{x}' = \arg\min_{\boldsymbol{x} \in (\mathcal{X}_L - \mathcal{X}_H)} f_L(\boldsymbol{x}) + \Psi_{\boldsymbol{w}}^{\boldsymbol{\psi}}(\boldsymbol{x})$
  $\gamma' = f_H(\boldsymbol{x}')$
  if $\gamma' < \tilde{\gamma}$ **then**
    $(\tilde{\boldsymbol{x}}, \tilde{\gamma}) = (\boldsymbol{x}', \gamma')$
  end if
  $\mathcal{X}_H, \boldsymbol{y} = \mathcal{X}_H \cup \{\boldsymbol{x}'\}, \boldsymbol{y} \cup \{\gamma' - f_L(\boldsymbol{x}')\}$
  $\mathbf{Z} = \mathbf{Z} \cup \{\psi_1(\boldsymbol{x}'), \psi_2(\boldsymbol{x}'), \ldots, \psi_k(\boldsymbol{x}')\}$
  $\boldsymbol{w} = (\mathbf{Z}^\top \mathbf{Z})^{-1} \mathbf{Z}^\top \boldsymbol{y}$
end for
return $(\tilde{\boldsymbol{x}}, \tilde{\gamma})$

$\boldsymbol{\psi} = \{\psi_1, \psi_2, \ldots, \psi_k\}$. We aggregate these base predictors as a final predictor using a simple weighted ensemble approach:

$$\Psi(\boldsymbol{x}) = \sum_{i=1}^{k} w_i \psi_i(\boldsymbol{x}) + b. \tag{12.5}$$

$\boldsymbol{w} = \{w_1, w_2, \ldots, w_k, b\}$ is a weight vector for the base predictors. $\mathcal{D} = \{(\boldsymbol{x}_1, y_1), (\boldsymbol{x}_2, y_2), \ldots\}$ is the raw regression training dataset. For each $\boldsymbol{x}_i$, we input it into the base predictors and obtain the weight ($\boldsymbol{w}$) training dataset: $\boldsymbol{Z} = \{z_1, z_2, \ldots\}$, where $z_i = \{\psi_1(\boldsymbol{x}_i), \psi_2(\boldsymbol{x}_i), \ldots, \psi_k(\boldsymbol{x}_i), 1\}$. $\boldsymbol{Z}$ is the feature matrix for training the weights. We rewrite $\boldsymbol{y} = \{y_1, y_2, \ldots\}$ as the learning target vector. By applying the mean square loss, the learning task can be presented as

$$\boldsymbol{w}^* = \arg\min_{\boldsymbol{w}} (\boldsymbol{y} - \boldsymbol{Z}\boldsymbol{w})^\top (\boldsymbol{y} - \boldsymbol{Z}\boldsymbol{w}). \tag{12.6}$$

The weights of the ensemble predictor have a closed-form solution: $\boldsymbol{w}^* = (\mathbf{Z}^\top \mathbf{Z})^{-1} \mathbf{Z}^\top \boldsymbol{y}$ when $\mathbf{Z}$ is a full-rank matrix. Thus, the process of training predictor Ψ has two steps: 1. using base predictors to predict $\mathcal{D}$ to obtain the weight training dataset $\mathbf{Z}$, and 2. calculating the weights with the above closed-form solution.

The linear ensemble approach of TSE is simple and easy to train. Another problem of TSE is how to obtain the base predictors.

Obtaining base predictors. The base predictors are also regression models. In this chapter, we decompose residual regression into middle-level problems. In hyper-parameter optimization, we naturally introduce a middle-fidelity evaluation function:

$$f_M(\delta) = \mathcal{L}(\delta, \mathcal{D}_{\text{sub}}^{r_M}, \mathcal{D}_{\text{valid}}), \tag{12.7}$$

where r_M is the subsample ratio and $0 < r_L < r_M \ll 1$. The middle-level problem is to train the regression predictor that estimates the regret function between f_M and f_L. Due to $r_M \ll 1$, the cost of f_M is much cheaper than the cost of f_H. We can obtain a large enough dataset for the middle-level problem. Thus, the middle-level predictor can be more accurate. If we need k base predictors in total, we should construct k middle-level regression problems: $\{(\mathcal{D}_{\text{sub1}}^{r_L}, \mathcal{D}_{\text{sub1}}^{r_M}), (\mathcal{D}_{\text{sub2}}^{r_L}, \mathcal{D}_{\text{sub2}}^{r_M}), \ldots, (\mathcal{D}_{\text{sub}k}^{r_L}, \mathcal{D}_{\text{sub}k}^{r_M})\}$. In addition, an extra $\mathcal{D}_{\text{sub}}^{r_L}$ is needed to construct the final regression problem $(\mathcal{D}_{\text{sub}}^{r_L}, \mathcal{D}_{\text{train}})$. Thus, a total of k $\mathcal{D}_{\text{sub}}^{r_M}$ and $k+1$ $\mathcal{D}_{\text{sub}}^{r_L}$ should be randomly sampled.

We implement the multi-fidelity DFO framework with TSE as Algorithm 12.2. The presented multi-fidelity optimization framework focuses on the evaluation phase of DFO. Only the evaluation phase is changed in optimization. Thus, this framework can be easily applied to any DFO method. Additionally, we present TSE to learn the residual predictor. Because high-fidelity evaluations are expensive, the training dataset is not large enough to learn an accurate predictor. TSE applies a linear combination of base predictors to simplify the regression model, and the base predictors prevent the combination predictor from starting from scratch. We obtain the base predictors by constructing middle-level regression problems related to the final regression problem. In hyper-parameter tuning problems, the middle-level regression problem estimates the simple regret between f_M and f_L. The base predictors are aligned by the learning model and can be transferred among different datasets. We focus only on the local trajectory of optimization to sample the regression training dataset. Thus, only a few instances are sufficient for the combination predictor to converge. With precise but cheap estimated evaluations as a substitute, optimization can explore more to find a better solution with an affordable cost.

12.3 Empirical Study

12.3.1 Experimental Setup

In the experiments of this chapter, we select LightGBM [7] as the learning algorithm. A total of 11 hyper-parameters of LightGBM constitute the search space, including learning rate, number of leaves, tree depth, number of rounds, etc. We select 12 datasets and show their details in Table 12.1. The dataset sizes range from thousands to 40 million. The subsampling ratio is set differently according to the dataset size. For small datasets with fewer than 100,000 instances, the low-fidelity subsampling

Table 12.1 Information about the datasets [3]. $|\mathcal{D}|$ is the number of instances in dataset $\mathcal{D}$. The validation datasets are constructed by sampling 10% of the instances from $\mathcal{D}^{\text{train}}$. r_L and r_M are the subsample ratios of $\mathcal{D}^{r_L}_{\text{sub}}$ and $\mathcal{D}^{r_M}_{\text{sub}}$

Dataset	$\lvert\mathcal{D}^{\text{train}}\rvert$	$\lvert\mathcal{D}^{\text{test}}\rvert$	r_L	r_M
Musk	4,991	2,083	0.05	0.2
HTRU2	14,318	3,580	0.05	0.2
Magic04	15,215	3,805	0.05	0.2
Credit	24,000	6,000	0.05	0.2
Adult	32,561	16,281	0.05	0.2
Sensorless	40,883	17,525	0.05	0.2
Connect	47,504	20,053	0.05	0.2
Miniboone	104,052	26,012	0.01	0.04
Airline	773,469	215,358	0.005	0.02
Higgs	10,000,000	1,000,000	0.001	0.004
MovieLens	16,000,210	4,000,053	0.001	0.004
Criteo	40,000,000	4,840,617	0.0005	0.002

ratio r_L is 0.05, and the middle-fidelity subsampling ratio r_M is 0.2. For large datasets, r_L and r_M depend on the dataset size.

We implement the multi-fidelity framework with TSE based on SRacos [2] and name it TSESRacos. The compared methods are as follows:

- TSETrans: The algorithm is the same as TSESRacos, but the base predictors are transferred from another dataset. In our experiments, the base predictors are all from the Miniboone dataset.
- RFSRacos: We replace the weighted ensemble of base predictors with a random forest regressor to obtain RFSRacos.
- LF-Only: We apply the raw SRacos to optimize the hyper-parameters using only the low-fidelity evaluation function.
- HF-Only: We apply the raw SRacos to optimize the hyper-parameters using only the high-fidelity evaluation function.

Because high-fidelity evaluations are accurate, HF-Only serves as the upper bound for all compared methods. However, due to the high time cost of high-fidelity evaluations, HF- Only optimizes the hyper-parameters on large datasets with only a few evaluations.

The evaluation criterion for the experiments is the AUC score. For multi-fidelity optimization methods (TSESRacos, TSETrans, and RFSRacos), we apply one high-fidelity evaluation for every 100 low-fidelity evaluations, i.e., $T_L = 100$. The total number of high-fidelity evaluations is set to 50, i.e., $T_H = 50$. Thus, the multi-fidelity optimization process has a total of 5,000 low-fidelity evaluations and 50 high-fidelity evaluations. For TSESRacos and TSETrans, there are 5 base predictors to construct the final regret predictor. For LF-Only, we set the total number of low-fidelity evaluations to 5,000, which is the same as for multi-fidelity optimization. For HF-ONLY, we set 5,000 evaluations only when the dataset size is less than 100,000 (Musk, HTRU2, Magic04, Credit, Adult, Sensorless, Connect, Miniboone). On large datasets (Airline, Higgs, MovieLens, Criteo), we early stop the optimization process when the time spent is more than that of TSETRANS (HF-ONLY*). For the huge datasets (MovieLens, Criteo), the middle-fidelity evaluation is hard to obtain. We can only transfer the base predictors from other tasks. Thus, we did not test TSESRACOS but only tested TSETRANS on these two datasets.

12.3.2 Empirical Analysis

We show the AUC scores and wall-clock running times of the compared methods on each dataset. For a fair comparison, we also include the time cost of training base predictors for TSESRACOS in the total wall-clock running time. Based on the experimental results, we can draw the following conclusions:

- Low-fidelity evaluation correction is necessary. From Table 12.2 , LF- ONLY usually achieves the best low-fidelity evaluation values. However, the corresponding high-fidelity evaluations are not good. Thus, it is necessary to correct the low-fidelity evaluation during optimization.
- Correction by a regression predictor is effective. From Table 12.2, the best AUC scores achieved by optimizations with correction (TSESRACOS, RFSRACOS, TSETRANS) are close to the upper bound score (HF- ONLY). They are much better than LF- ONLY on most datasets.
- The TSE regressor converges fast. In Figure 12.1, we compare the regression error of the TSE regressor to that of the random forest regressor. At the beginning (when the regression training dataset has only one instance), TSE has a larger error. However, the error of TSE decreases quickly when the dataset size exceeds 5. In particular, the error variance of the TSE regressor is smaller than that of the random forest regressor, indicating that the TSE regressor has good stability.
- Base predictors have the ability to transfer. In Table 12.2, except for TSESRACOS, TSETRANS ranks first 10 times out of 11 datasets. Compared to TSESRACOS, TSETRANS achieves similar optimization performance but spends less time because it does not require a base predictor training phase. This verifies that base predictors can be easily transferred to other datasets. This is meaningful for huge datasets where training base predictors is challenging.
- Multi-fidelity optimization with TSE is effective. In Table 12.2, TSESRACOS outperforms others in most cases (ranking first 8 times out of 10 datasets). Compared to LF- ONLY and HF- ONLY, the performance of TSESRACOS is close to HF-ONLY, while the time cost is close to LF- ONLY. This verifies that multi-fidelity optimization with TSE can significantly improve optimization performance with an acceptable extra time cost.

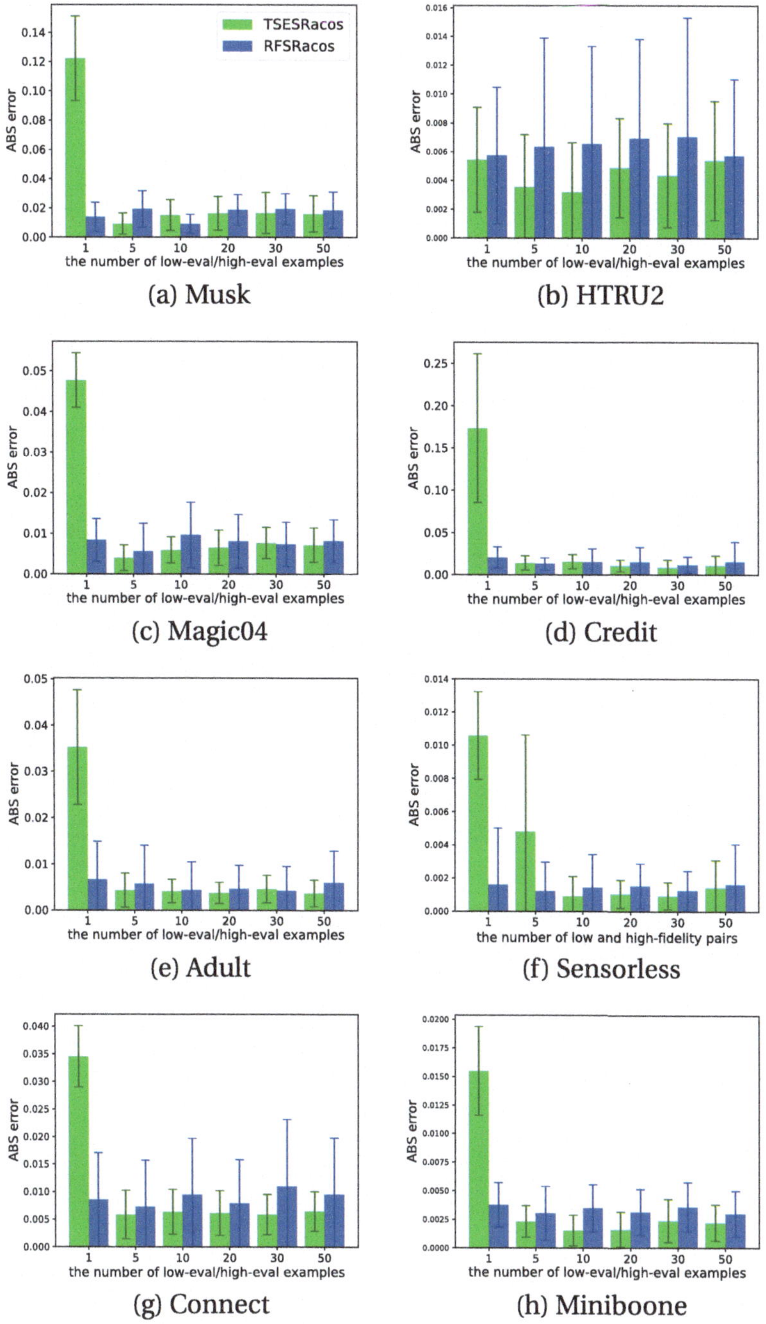

(a) Musk (b) HTRU2

(c) Magic04 (d) Credit

(e) Adult (f) Sensorless

(g) Connect (h) Miniboone

Fig. 12.1 Histograms of the mean regression prediction error $|f_L + \Psi - f_H|$ at each training step [3]. It only compares the prediction error of TSESRacos (green) and RFSRacos (blue). The X-axis represents the number of instances in the regression training dataset

Table 12.2 AUC performance and wall-clock time of the compared methods [3]. LF-Eval and HF-Eval represent the best low and high-fidelity evaluation values of the searched hyper-parameters. Test represents the generalization AUC score of the searched hyper-parameters. The bold numbers in each column indicate the best AUC score

Dataset	Method	LF-Eval	HF-Eval	Test	Time	Dataset	Method	LF-Eval	HF-Eval	Test	Time
Musk	TseSRacos	0.9018	**0.9991**	0.9977	0:07:31	HTRU2	TseSRacos	0.9733	**0.9841**	0.9632	0:02:44
	TseTrans	0.9204	**0.9991**	**0.9985**	0:07:16		TseTrans	0.9650	0.9758	**0.9636**	0:01:41
	RFSRacos	0.9220	0.9990	0.9980	0:06:40		RFSRacos	0.9773	0.9814	0.9616	0:01:27
	LF- Only	0.9294	0.9989	0.9974	0:05:46		LF-Only	0.9750	0.9791	0.9613	0:02:11
	HF- Only	–	1.0000	0.9978	1:49:08		HF-Only	–	0.9871	0.9645	0:08:48
Magic04	TseSRacos	0.8859	**0.9446**	**0.9236**	0:04:40	Credit	TseSRacos	0.6407	**0.7451**	**0.7612**	0:03:46
	TseTrans	0.9013	0.9438	0.9227	0:03:04		TseTrans	0.6654	0.7432	0.7531	0:01:36
	RFSRacos	0.8994	0.9387	0.9225	0:02:16		RFSRacos	0.6889	0.7404	0.7579	0:01:21
	LF-Only	0.9092	0.9296	0.9201	0:02:45		LF-Only	0.7270	0.7324	0.7531	0:00:59
	HF-Only	–	0.9495	0.9203	0:20:06		HF-Only	–	0.7554	0.7643	0:04:26
Adult	TseSRacos	0.8961	**0.9261**	**0.9219**	0:04:20	Sensorless	TseSRacos	0.9974	**0.9999**	**0.9999**	0:33:29
	TseTrans	0.8896	0.9224	0.9206	0:02:17		TseTrans	0.9973	**0.9999**	**0.9999**	0:21:33
	RFSRacos	0.9086	0.9190	0.9181	0:02:37		RFSRacos	0.9978	**0.9999**	0.9998	0:54:30
	LF-Only	0.9070	0.9157	0.9156	0:02:37		LF-Only	0.9973	0.9997	0.9997	0:19:38
	HF-Only	–	0.9281	0.9234	0:26:02		HF-Only	–	0.9999	0.9999	2:23:44
Connect	TseSRacos	0.8604	0.9318	**0.9374**	0:11:02	Miniboone	TseSRacos	0.9664	**0.9789**	**0.9785**	0:27:53
	TseTrans	0.8650	**0.9319**	0.9353	0:11:14		TseTrans	–	–	–	–
	RFSRacos	0.8630	0.9284	0.9330	0:09:15		RFSRacos	0.9674	0.9787	0.9781	0:12:44
	LF-Only	0.8684	0.9219	0.9272	0:10:32		LF-Only	0.9694	0.9779	0.9771	0:14:54
	HF-Only	–	0.9367	0.9404	1:20:28		HF-Only	-	0.9814	0.9797	0:51:00
Airline	TseSRacos	0.6392	**0.6801**	**0.8893**	0:44:06	Higgs	TseSRacos	0.7743	0.8037	0.8023	14:09:18
	TseTrans	0.6462	0.6674	0.8696	0:40:31		TseTrans	0.7770	**0.8046**	**0.8044**	11:37:50
	RFSRacos	0.6519	0.6674	0.8762	0:35:55		RFSRacos	0.7847	0.8025	0.8035	12:57:22
	LF-Only	0.6566	0.6600	0.8693	0:41:29		LF-Only	0.7872	0.7991	0.7988	8:53:33
	HF-Only*	–	0.6900	0.8961	2:00:00		HF-Only*	–	0.8145	0.8140	45:00:00
MovieLens	TseTrans	0.6344	**0.6682**	0.6476	11:53:56	Criteo	TseTrans	0.7258	**0.7513**	**0.7496**	62:00:25
	RFSRacos	0.6444	0.6543	**0.6591**	11:35:42		RFSRacos	0.7289	0.7454	**0.7496**	65:41:30
	LF-Only	0.6443	0.6477	0.6361	11:10:26		LF-Only	0.7298	0.7480	0.7480	60:52:23
	HF-Only*	–	0.6767	0.6591	36:00:00		HF-Only*	–	0.7652	0.7584	180:00:00

12.4 Summary

This chapter applies multi-fidelity optimization to address the expensive evaluation issue in hyper-parameter optimization. In hyper-parameter optimization, low-fidelity evaluations can be obtained by validating the hyper-parameter configuration on a subset of the dataset. These evaluations are cheap but significantly biased. Previous works on multi-fidelity optimization are often based on specific methods. In this chapter, we present a general multi-fidelity optimization framework based on DFO. A correction predictor is trained to estimate the residual between high and low-fidelity evaluations. We can optimize according to the corrected low-fidelity evaluations to reduce the time cost. However, high-fidelity evaluations are still hard to obtain, and the regression dataset is too small to train an accurate predictor. We present Transfer Series Expansion (TSE) to address this issue. TSE linearly combines pre-trained base predictors. Additionally, base predictors are trained on middle-level regression problems, for which training datasets are easier to obtain. Experiments on LightGBM hyper-parameter optimization tasks verify that multi-fidelity optimization with TSE can effectively accelerate the optimization process.

References

1. M. G. Fernández-Godino, C. Park, N.-H. Kim, and R. T. Haftka. Review of multi-fidelity models. *arXiv preprint* arXiv:1609.07196, 2016
2. Y. Hu, H. Qian, and Y. Yu. Sequential classification-based optimization for direct policy search. In *Proceedings of the 31st AAAI Conference on Artificial Intelligence*, pages 2029–2035, San Francisco, CA, 2017
3. Y.-Q. Hu, Y. Yu, W.-W. Tu, Q. Yang, Y. Chen, and W. Dai. Multi-fidelity automatic hyper-parameter tuning via transfer series expansion. In *Proceedings of the 33rd AAAI Conference on Artificial Intelligence (AAAI'19)*, Honolulu, HI, 2019
4. Huang D, Allen TT, Notz WI, Miller RA (2006) Sequential kriging optimization using multiple-fidelity evaluations. Structural and Multidisciplinary Optimization 32(5):369–382
5. K. Kandasamy, G. Dasarathy, J. B. Oliva, J. Schneider, and B. Poczos. Multi-fidelity gaussian process bandit optimisation. *arXiv preprint* arXiv:1603.06288, 2016
6. K. Kandasamy, G. Dasarathy, J. Schneider, and B. Poczos. Multi-fidelity bayesian optimisation with continuous approximations. *arXiv preprint* arXiv:1703.06240, 2017
7. G. Ke, Q. Meng, T. Finley, T. Wang, W. Chen, W. Ma, Q. Ye, and T.-Y. Liu. Lightgbm: A highly efficient gradient boosting decision tree. In *Advances in Neural Information Processing Systems*, pages 3146–3154, 2017
8. March A, Willcox K (2012) Provably convergent multifidelity optimization algorithm not requiring high-fidelity derivatives. AIAA journal 50(5):1079–1089
9. R. Sen, K. Kandasamy, and S. Shakkottai. Multi-fidelity black-box optimization with hierarchical partitions. In *Proceedings of the 35th International Conference on Machine Learning (ICML'18)*, pages 4545–4554, 2018

Chapter 13
Stepwise Optimization: Cascaded Algorithm Selection

Abstract This chapter introduces a stepwise optimization approach for algorithm selection in Automatic Machine Learning (AutoML). Traditional methods like Combined Algorithm Selection and Hyper-parameter optimization (CASH) often suffer from inefficiency due to the large and redundant search space. To address this, the chapter proposes a cascaded algorithm selection framework, which separates the process into two levels: hyper-parameter optimization for individual algorithms and a resource allocation strategy at the upper level. The upper level is formulated as a multi-armed bandit problem, where each arm represents a hyper-parameter optimization process. The chapter introduces the Extreme-Region Upper Confidence Bound (ER-UCB) strategy, designed to maximize the final feedback by focusing on the arm with the largest extreme region. Theoretical analysis and empirical studies on synthetic and real-world AutoML tasks demonstrate the effectiveness of ER-UCB in improving algorithm selection efficiency and performance. The results highlight the importance of stepwise optimization in reducing redundancy and enhancing AutoML outcomes.

Algorithm selection [1, 4, 5] and hyper-parameter optimization [9, 10], introduced in the previous chapters, are two core tasks of Automatic Machine Learning (AutoML) [17]. While hyper-parameter optimization has been extensively studied by researchers recently [2, 3, 8, 14], there are fewer works focusing on algorithm selection. Some approaches solve algorithm selection by combining it with hyper-parameter optimization, such as the *Combined Algorithm Selection and Hyper-parameter optimization* (CASH) method, which searches within a joint hyper-parameter space of all candidate algorithms. However, the search result is the best hyper-parameter configuration of the best algorithm, and the hyper-parameter spaces constructed by other algorithms are redundant. The optimization solver, typically a derivative-free optimization (DFO) method such as Bayesian optimization [12, 13], is sensitive to dimensionality. The large and redundant search space hinders the optimization solver from reaching its full potential and obtaining a good result.

To address this issue, we present a stepwise optimization approach to solve algorithm selection: cascaded algorithm selection [11]. Cascaded algorithm selection consists of a two-level process. The lower level is the hyper-parameter optimization process for each algorithm. However, the optimization resources, such as running time and sample budget, are usually limited. The upper level is a strategy to solve

Y. Yu et al., *Derivative-Free Optimization*, Machine Learning: Foundations, Methodologies, and Applications, https://doi.org/10.1007/978-981-96-5929-6_13

how to reasonably allocate the optimization resources to each lower-level process. We formulate this allocation problem as a multi-armed bandit problem and present an *Extreme-Region Upper Confidence Bound* (ER-UCB) strategy [11]. Theoretical analysis under ideal assumptions and empirical studies on real algorithm selection tasks show that ER-UCB is an effective strategy for cascaded algorithm selection.

13.1 Stepwise Optimization with Algorithm Selection

Let $\boldsymbol{\mathcal{A}} = \{\mathcal{A}_1, \mathcal{A}_2, \ldots, \mathcal{A}_K\}$ denote a set of K candidate learning algorithms. For each algorithm $\mathcal{A}_i$, let $\boldsymbol{\delta}_i \in \boldsymbol{\Delta}_i$ denote a hyper-parameter configuration, where $\boldsymbol{\Delta}_i$ is the hyper-parameter space of $\mathcal{A}_i$. Given a training dataset $\mathcal{D}_{\text{train}}$, a testing dataset $\mathcal{D}_{\text{test}}$, and an evaluation criterion $f(\cdot)$, an AutoML task can be formulated as

$$\mathcal{A}_{i^*}^{\boldsymbol{\delta}_{i^*}^*} = \mathop{\arg\max}_{i \in \{1,2,\ldots,K\}} \mathop{\arg\max}_{\boldsymbol{\delta}_i \in \boldsymbol{\Delta}_i} \frac{1}{k} \sum_{j=1}^{k} f(\mathcal{A}_i^{\boldsymbol{\delta}_i}, \mathcal{D}_{\text{train}}^j, \mathcal{D}_{\text{valid}}^j), \tag{13.1}$$

where $\mathcal{D}_{\text{valid}}^j \subset \mathcal{D}_{\text{train}}$ and $\mathcal{D}_{\text{train}}^j = \mathcal{D}_{\text{train}} \setminus \mathcal{D}_{\text{valid}}^j$. In this chapter, we consider a maximization problem, where $f(\cdot)$ is an evaluation criterion such as accuracy or AUC score, and a larger value is better.

We note that Eq. (13.1) contains two cascaded optimization processes. Thus, we can rewrite Eq. (13.1) as follows:

$$\mathcal{A}_{i^*}^{\boldsymbol{\delta}_{i^*}^*} = \mathop{\arg\max}_{i \in \{1,2,\ldots,K\}} \frac{1}{k} \sum_{j=1}^{k} f(\mathcal{A}_i^{\boldsymbol{\delta}_i^*}, \mathcal{D}_{\text{train}}^j, \mathcal{D}_{\text{valid}}^j),$$

$$\text{where } \boldsymbol{\delta}_i^* = \mathop{\arg\max}_{\boldsymbol{\delta}_i \in \boldsymbol{\Delta}_i} \frac{1}{k} \sum_{j=1}^{k} f(\mathcal{A}_i^{\boldsymbol{\delta}_i}, \mathcal{D}_{\text{train}}^j, \mathcal{D}_{\text{valid}}^j).$$

The inner part corresponds to the lower-level optimization process, which is the hyper-parameter optimization process. We assume that the best hyper-parameters $\boldsymbol{\delta}_i^*$ are available for every algorithm $\mathcal{A}_i$. The upper-level optimization process is straightforward, as it only needs to select the algorithm with the highest evaluation value. However, in reality, the cost for the lower-level hyper-parameter optimization to obtain $\boldsymbol{\delta}_i^*$ is too high to be acceptable. Thus, the direct algorithm selection strategy is not applicable. We tackle this issue by a **stepwise** optimization approach, i.e., cascaded algorithm selection.

The lower level of cascaded algorithm selection. The lower level contains a set of hyper-parameter optimization processes, one for each algorithm $\mathcal{A}_i$. DFO is the solver for this task. Current DFO methods share a similar optimization framework, i.e., a `Sample&Evaluate` loop. `Sample` denotes a sampling step that builds a model based on previous samples and generates a new sample. `Evaluate` denotes an evaluation step that evaluates the new sample, and the evaluation result is the feedback to the DFO method. By repeating this `Sample&Evaluate` loop, the DFO method improves the search performance iteratively. We also note that DFO has a stepwise optimization process. The optimization loop can be paused at any iteration and resumed if necessary. Compared to searching on the joint hyper-parameter space (CASH), the optimization process at the lower level of cascaded algorithm selection focuses only on the hyper-parameter space of each algorithm. All search dimensions are active, mitigating the issue of redundant search. Furthermore, the dimensionality of each optimization process is small, allowing the DFO method to obtain better performance more easily.

The upper level of cascaded algorithm selection. The upper level is a resource allocation strategy for the hyper-parameter optimization processes at the lower level. With K learning algorithms, we have a total of K hyper-parameter optimization processes at the lower level. Each hyper-parameter optimization process requires resources to complete the search. For example, the evaluation criterion for hyper-parameter optimization is usually a k-fold cross-validation process, which consumes time and computational resources to obtain the evaluation value. In an AutoML system, the total resource is usually pre-defined. When the resource is exhausted, the AutoML system stops and returns the best configuration found so far. The goal of the upper level is to reasonably allocate the resource to the hyper-parameter optimization processes at the lower level. Since DFO is the solver for hyper-parameter optimization tasks and has a stepwise framework that can be paused and resumed at any iteration, the upper level can be seen as a multi-armed bandit problem. An arm represents a hyper-parameter optimization process of an algorithm. The action of pulling an arm is to select the corresponding hyper-parameter optimization process and run it for one iteration. The key to the upper level is the arm selection strategy under the multi-armed bandit formulation.

Multi-armed bandit for cascaded algorithm selection. Figure 13.1 illustrates the multi-armed bandit formulation of cascaded algorithm selection. The right-hand part (the arm) is the stepwise DFO process. The action of pulling an arm is to resume the `Sample` step to generate a hyper-parameter configuration and evaluate this configuration. The evaluation value is the feedback of pulling the arm. Since the frequently used DFO methods are stochastic, the hyper-parameter optimization process is a stochastic process. For example, if random search is employed as the hyper-parameter optimization solver, the feedback distribution is the performance distribution of the algorithm in the defined hyper-parameter space. Usually, the feedback distribution is unknown to us. The key to the arm selection strategy is to balance exploration and exploitation when pulling arms.

13.2 The ER-UCB Algorithm

In the multi-armed bandit formulation, the key is the arm selection strategy. We present the *Extreme-Region Upper Confidence Bound* (ER-UCB) strategy for

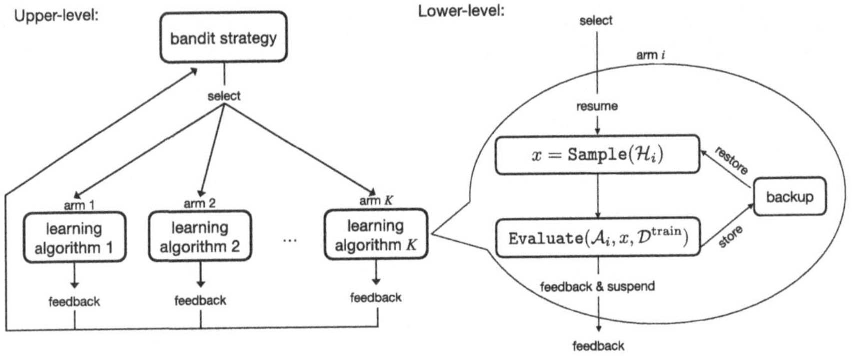

Fig. 13.1 Illustration of the cascaded algorithm selection framework [11]. The lower level (right-hand side) is a stepwise hyper-parameter optimization process that contains `Sample` and `Evaluate` steps. The upper level (left-hand side) is an arm selection strategy that decides which arm should be pulled in the next iteration

cascaded algorithm selection. We assume that we have K arms and a total of n trials for selecting arms. For the i-th arm at the t-th trial, let $X_{i,t} \sim \mathfrak{D}_i(\mu_{X_{i,t}}, \sigma^2_{X_{i,t}})$ denote the random variable of the feedback, where $\mathfrak{D}_i$ denotes the underlying distribution with the expectation $\mathbb{E}[X_{i,t}] = \mu_{X_{i,t}}$ and the variance $\mathbb{D}[X_{i,t}] = \sigma^2_{X_{i,t}}$. In the well-studied multi-armed bandit problem, the target is to maximize the long-term accumulated feedback. The *Upper Confidence Bound* (UCB) strategy is a widely used arm selection strategy in this setting, which aims to find the arm with the maximal feedback expectation. However, cascaded algorithm selection aims to find the best algorithm and its best hyper-parameters. In the multi-armed bandit setting, the strategy should maximize the single feedback. Thus, the classic UCB strategy is inappropriate for cascaded algorithm selection. We present the extreme region of the feedback distribution to meet the target of cascaded algorithm selection.

13.2.1 Extreme Region Target and Extreme Region Regret

Definition 13.1 (*Extreme Region*) Assuming $X \sim \mathfrak{D}$, where X is a random variable and $\mathfrak{D}$ is the corresponding distribution, with a constant real number ρ, the extreme region is

$$\text{ER}(\mathfrak{D}) = \Pr[X > \rho]. \tag{13.2}$$

The extreme region (Definition 13.1) is the area under the probability density function of the feedback distribution to the right of a constant value ρ. Intuitively, the extreme region indicates the probability of obtaining a sample larger than ρ when sampling from the feedback distribution. Given a fixed ρ, we want to find the arm with the largest extreme region among all arms, as it is most likely to obtain a sample larger than ρ by pulling this arm. Thus, the extreme region satisfies the requirement of cascaded algorithm selection.

Figure 13.2 shows an example of the extreme region for two synthetic arms: Gaussian distributions with $(\mu_1 = 0.5, \sigma_1^2 = 0.15^2)$ and $(\mu_2 = 0.6, \sigma_2^2 = 0.05^2)$. Although arm 2 has a larger expectation than arm 1 ($\mu_1 < \mu_2$), we still select arm 1 due to its larger extreme region ($\mathrm{ER}(\mathfrak{D}_1) > \mathrm{ER}(\mathfrak{D}_2)$).

Definition 13.2 (*Extreme Region Regret*) Let i^* be the index of the best arm. With a fixed ρ and $i \in \{1, 2, \ldots, K\}$, let $p_i = \Pr[X_i > \rho]$ and $p^* = \Pr[X_{i^*} > \rho]$. With n total trials, a strategy selects the I_tth arm at the tth trial, where $t \in \{1, 2, \ldots, n\}$. The extreme region regret of this strategy is

$$R_n = np^* - \mathbb{E}\sum_{t=1}^{n} p_{I_t}. \tag{13.3}$$

We define the *extreme region regret* (Definition 13.2) to measure the gap between the real best strategy and the practical strategy. Intuitively, the extreme region regret is the difference in the number of events where a sample is larger than ρ between the real best strategy and the practical strategy.

13.2.2 ER-UCB on Stationary Distributions

We first consider the scenario where the feedback distributions are stationary, which corresponds to the case where the hyper-parameter optimization method is random search. The ER-UCB-S strategy for stationary distributions is as follows:

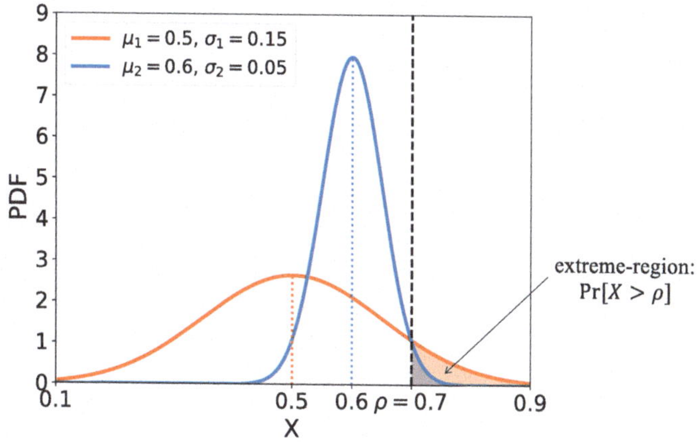

Fig. 13.2 Illustration of the extreme region on the probability density function (PDF) of a distribution [11]. We assume that the feedback distributions of two arms follow the Gaussian distributions: $\mathcal{G}_1(0.5, 0.15^2)$ and $\mathcal{G}_2(0.6, 0.05^2)$. With a constant ρ, we define the extreme region as the probability $\Pr[X > \rho]$. In this figure, we set $\rho = 0.7$. The shaded areas under the PDF lines are the extreme regions of the two arms

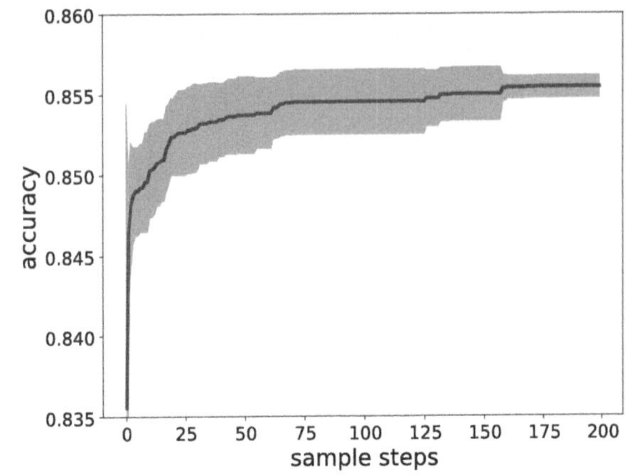

Fig. 13.3 Illustration of the convergence curve of the hyper-parameter tuning process [11]. We apply a DFO method to optimize the hyper-parameters of a decision tree on the Adult dataset. The total sample budget is 200. We run this experiment 10 independent times. The average convergence curve is plotted

$$
\begin{aligned}
I_t &= \operatorname*{arg\,max}_{i \in \{1,2,\ldots,K\}} \; \gamma \hat{\mu}_i(t) + g_i(t), \text{ where,} \qquad (13.4)\\
\hat{\mu}_i(t) &= \hat{\mu}_{Y_i}^{T_i(t)} + \sqrt{\frac{1}{\theta} \hat{\mu}_{Y_i^2}^{T_i(t)}},\\
g_i(t) &= \sqrt{\frac{2 \ln t}{T_i(t)}} + \sqrt{\frac{1}{\theta}} \sqrt{\frac{2 \ln t}{T_i(t)}}.
\end{aligned}
$$

Here, $\hat{\mu}_i(t)$ is the exploitation term, which is the estimated extreme region of the ith arm based on the observed feedbacks. $g_i(t)$ is the exploration term, which reflects the uncertainty in the extreme region estimation. γ is a hyper-parameter for balancing exploration and exploitation. $T_i(t)$ is the number of times the ith arm has been pulled up to the tth trial. θ is a hyper-parameter related to the size of the extreme region.

13.2.3 ER-UCB on Non-stationary Distributions

In the non-stationary setting, the feedback distributions change as the optimization progresses, which corresponds to the case where the hyper-parameter optimization method is an advanced local search method. Figure 13.3 illustrates the typical convergence curve of hyper-parameter optimization, which increases quickly at the beginning and slows down at the end. We assume that the curve can be represented by a parameterized function $\phi(at + b)$, where a and b are undetermined coefficients.

The ER-UCB-N strategy for non-stationary distributions is as follows:

$$I_t = \underset{i \in \{1,2,\ldots,K\}}{\arg\max} \ \gamma \hat{\mu}_i(t) + g_i(t), \text{ where,} \tag{13.5}$$

$$\hat{\mu}_i(t) = \hat{a}_i t + \hat{b}_i + \sqrt{\frac{1}{\theta}\hat{\sigma}^2_{Z_{i,t}}},$$

$$g_i(t) = \Delta_{T_i(t)}(t) + \sqrt{(\Delta_{T_i(t)}(T_i(t)) + 1)\sqrt{\frac{\alpha \ln t}{2T_i(t)}}}.$$

Here, $\hat{\mu}_i(t)$ is the estimated extreme region of the ith arm, which is modeled as a linear function of t with coefficients $\hat{a}_i$ and $\hat{b}_i$. $\hat{\sigma}^2_{Z_{i,t}}$ is the estimated variance of the transformed feedback $Z_{i,t} = \phi^{-1}(X_{i,t})$. $g_i(t)$ is the exploration term, which depends on the number of times the ith arm has been pulled ($T_i(t)$) and the total number of trials (t). $\Delta_{T_i(t)}(t)$ is a function that captures the uncertainty in the extreme region estimation. γ and α are hyper-parameters for balancing exploration and exploitation.

13.2.4 Theoretical Results

We provide theoretical guarantees for the ER-UCB-S and ER-UCB-N algorithms in terms of the extreme region regret. Intuitively, the theorems state that the extreme region regret grows sublinearly with the number of trials n, which means that the algorithms converge to the optimal strategy as n increases.

Theorem 13.1 *(Extreme Region Regret of ER-UCB-S) Under certain assumptions, the extreme region regret of the ER-UCB-S algorithm satisfies:*

$$R_n \le \sum_{i:\Gamma_i>0} \Theta_i \left(\frac{32 \ln n}{\Gamma_i^4/(1+\theta^{-1})^4} + 3 \right), \tag{13.6}$$

where Γ_i and Θ_i are constants related to the extreme regions of the arms.

Theorem 13.2 *(Extreme Region Regret of ER-UCB-N) Under certain assumptions, the extreme region regret of the ER-UCB-N algorithm satisfies:*

$$R_n \le \sum_{i:\Gamma_i>0} \Theta_i(\kappa(n) + \frac{12((1+n)^{1-\alpha} - 1)}{1-\alpha} + 1), \tag{13.7}$$

where $\kappa(n)$ is a sublinear function of n, and α, Γ_i, Θ_i are constants related to the extreme regions of the arms.

We ignore the proofs in this section. Readers interested in the detailed proofs of these theorems are referred to [11].

13.3 Empirical Study

We design synthetic and AutoML tasks to investigate the ER-UCB algorithm (ER-UCB-S and ER-UCB-N). We choose several bandit algorithms as the compared methods, including extreme bandits [7] (Extreme), the classic UCB [6] (C-UCB), ε-Greedy [15], Softmax strategy [16], successive halving (S-Halving), UCB-E, and random strategy. For AutoML tasks, we choose the state-of-the-art algorithm selection method, AutoSklearn, as a compared method.

13.3.1 Synthetic Tasks

We design synthetic tasks with stationary arms and non-stationary arms separately. Due to the available feedback distribution of the arm, we can conveniently investigate the presented methods. We present the experiment details and the results.

Stationary Setting

We set 7 arms in this multi-armed bandit task. The feedback distribution of each arm is a Gaussian distribution. The expectations and variances of the 7 arms are $\mathcal{G}_1(0.64, 0.05^2)$, $\mathcal{G}_2(0.64, 0.01^2)$, $\mathcal{G}_3(0.65, 0.03^2)$, $\mathcal{G}_4(0.65, 0.02^2)$, $\mathcal{G}_5(0.68, 0.01^2)$, $\mathcal{G}_6(0.68, 0.02^2)$, and $\mathcal{G}_7(0.69, 0.01^2)$. According to the extreme region definition, $\mathcal{G}_1$ is the best arm in this bandit, i.e., $i^* = 1$. Thus, a good strategy should allocate trials to $\mathcal{G}_1$ as many as possible. We define the exploitation rate $R_i^{\text{eoi}} = \frac{T_i(n)}{n}$ to measure the percent of trials allocated to the ith arm. For the best i^*th arm, a larger $R_{i^*}^{\text{eoi}}$ is better. For ER-UCB-S, we set $\theta = 0.01$, $\beta = 0.66$, and $\gamma = 20$. For ER-UCB-N, we set $\phi(x) = x$, $\theta = 0.01$, and $\gamma = 20$. We set the total number of trials to $n = 1000$ for all compared methods. All experiments are run for 3 independent times. The average performance is reported.

Table 13.1 shows the average performance of all compared methods. Some criteria are proposed to investigate the effectiveness of the methods: $\bar{X}^*$ is the best average final feedback among 3 runs. $\arg\max_i X_i^*$ shows from which arm X^* is obtained. We report R_1^{eoi} and R_7^{eoi} in this task. $\mathcal{G}_1$ is the best arm. $\mathcal{G}_7$ has the largest expectation, which is easy to mislead arm selection strategies. $\arg\max_i R_i^{\text{eoi}}$ shows which arm is most frequently pulled. According to the results, we can draw the following conclusions:

- It is most likely to obtain an extreme value by pulling $\mathcal{G}_1$. The random strategy uniformly allocates trials to each arm. However, X^* always occurs at $\mathcal{G}_1$, i.e., $\{1, 1, 1\} = \arg\max_i X_i^*$ in the Random row of Table 13.1.
- The methods that maximize average expectation are easily misled by $\mathcal{G}_7$, such as C-UCB, ε-greedy, and UCB-E, which fail to allocate trials to the correct arm ($\mathcal{G}_1$).
- The methods that maximize extreme values successfully find $\mathcal{G}_1$. ER-UCB-S, ER-UCB-N, and Extreme achieve the result of $\arg\max_i R_i^{\text{eoi}} = \{1, 1, 1\}$.

Table 13.1 Average performance on the stationary setting [11]. $\bar{X}^*$ denotes the best average final feedback from 3 runs. $\arg\max_i X_i^*$ denotes the arm index from which X^* is sampled. $\arg\max_i R_i^{\text{eoi}}$ denotes the arm index to which the strategy allocates the most trials. R_1^{eoi} and R_7^{eoi} are exploitation rates of arms 1 and 7. The bold number is the best performance in its column

Method	$\bar{X}^*$	$\arg\max_i X_i^*$	$\arg\max_i R_i^{\text{eoi}}$	R_1^{eoi}	R_7^{eoi}
ER-UCB-N	0.7925	1, 1, 1	1, 1, 1	0.7463	0.0416
ER-UCB-S	**0.8079**	1, 1, 1	1, 1, 1	**0.8613**	0.0100
Extreme	0.7626	1, 1, 1	1, 1, 1	0.1660	0.1390
C-UCB	0.7353	3, 6, 6	7, 7, 7	0.0103	0.6996
Softmax	0.7860	1, 1, 1	7, 6, 3	0.1446	0.1520
ε-Greedy	0.7286	7, 6, 6	7, 7, 7	0.0123	0.8853
S-Halving	0.7675	3, 1, 1	3, 1, 1	0.3819	0.0472
UCB-E	0.7506	1, 6, 1	7, 7, 7	0.0783	0.2596
Random	0.7650	1, 1, 1	6, 2, 5	0.1396	0.1446

- The ER-UCB algorithms show better efficiency than Extreme. R_1^{eoi} of Extreme is only 0.166, indicating that Extreme spends more trials on exploration. R_1^{eoi} of ER-UCB-S and ER-UCB-N are 0.8613 and 0.7463, indicating that ER-UCB can quickly lock onto $\mathcal{G}_1$ and allocate trials to it.
- By setting $\phi(x) = x$, ER-UCB-N is also effective in the stationary setting. ER-UCB-N receives high R_1^{eoi} and correctly finds that $\mathcal{G}_1$ is the best arm within three repeated runs.
- The regret study shows that ER-UCB is the best arm selection strategy among the compared methods. In Fig. 13.4.1, the lines of ER-UCB-N (green line) and ER-UCB-S (blue line) approach the expectation of the best strategy (dashed red line) faster than others.

Non-stationary Setting

We design the non-stationary setting as follows. We set $\phi(at+b) = 2/(1+\exp(-0.005(at+b))) - 1$ to simulate the convergence curve. The arm is represented by $\mathcal{A}_i(a, b, \sigma_i^2)$. We set a total of 6 arms: $\mathcal{A}_1(\frac{5}{9}, 50, 0.03^2)$, $\mathcal{A}_2(\frac{5}{9}, 60, 0.01^2)$, $\mathcal{A}_3(\frac{2}{7}, 120, 0.03^2)$, $\mathcal{A}_4(\frac{2}{7}, 60, 0.02^2)$, $\mathcal{A}_5(\frac{1}{6}, 50, 0.02^2)$, and $\mathcal{A}_6(\frac{1}{6}, 60, 0.015^2)$. $\mathcal{A}_3$ has the largest expectation at the beginning. But $\mathcal{A}_1$ and $\mathcal{A}_2$ have larger long-term expectations. Although $\mathcal{A}_2$ has a slightly larger expectation than $\mathcal{A}_1$, due to the larger σ^2, $\mathcal{A}_1$ is the best arm. We run all compared methods for 3 independent times. The total number of trials is set to $n = 1000$. For ER-UCB-N, we set $\phi(x) = 2/(1+\exp(-0.005x)) - 1$, $\theta = 0.01$, and $\gamma = 4.5$. For ER-UCB-S, we set $\theta = 0.01$, $\beta = 0.85$, and $\gamma = 20$ (Table 13.2 and Fig. 13.5).

Table 13.2 Performance of the compared methods on the non-stationary setting [11]. Please refer to Table 13.1 for the column instructions

Method	$\bar{X}^*$	$\arg\max_i X_i^*$	$\arg\max_i R_i^{eoi}$	R_1^{eoi}	R_3^{eoi}
ER-UCB-N	**0.9055**	1, 1, 1	1, 1, 1	**0.7799**	0.0600
ER-UCB-S	0.3508	1, 1, 1	5, 5, 5	0.1460	0.0023
Extreme	0.5689	1, 1, 1	1, 1, 1	0.3150	0.1370
C-UCB	0.8465	3, 3, 3	3, 3, 3	0.0010	0.9946
Softmax	0.4664	3, 3, 3	3, 3, 3	0.1630	0.1916
ε-Greedy	0.7917	3, 3, 3	3, 3, 3	0.0176	0.9136
S-Halving	0.6336	3, 3, 3	3, 3, 3	0.1667	0.5010
UCB-E	0.8039	3, 3, 3	3, 3, 3	0.0130	0.9303
Random	0.4401	3, 3, 3	5, 3, 2	0.1673	0.1663

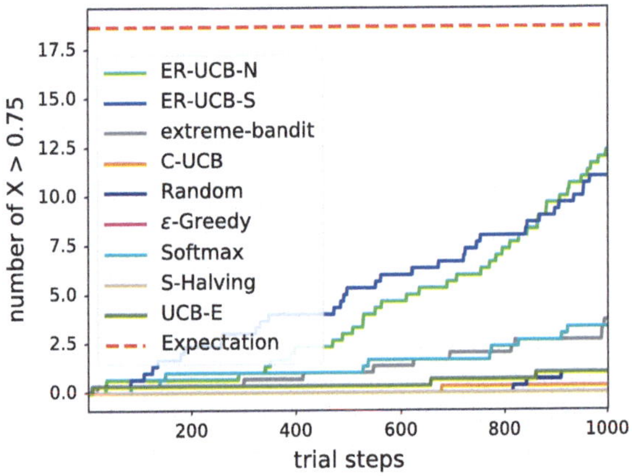

Fig. 13.4 Illustration of the regret curve on the stationary setting of synthetic tasks [11]. The red dashed line is the expectation of the best arm selection strategy

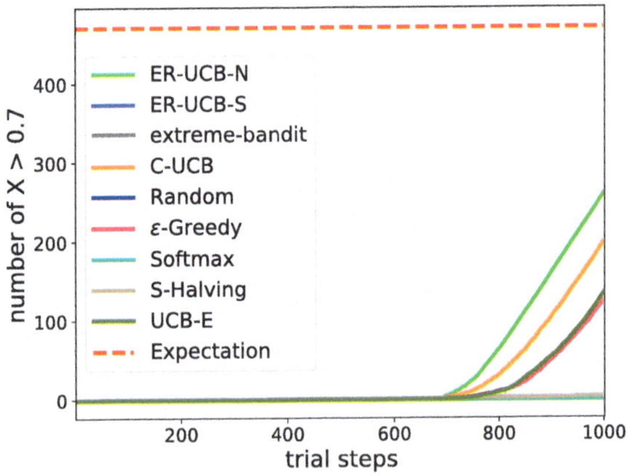

Fig. 13.5 Illustration of the regret curve on the non-stationary setting of synthetic tasks [11]. The red dashed line is the expectation of the best arm selection strategy

According to the results, we can draw the following conclusions:

- $\mathcal{A}_1$ is the best arm in this setting. In ER-UCB-N, the best $\bar{X}^* = 0.9055$ is obtained on $\mathcal{A}_1$ with $R_1^{\text{eoi}} = 0.7799$. In C-UCB, the best $\bar{X}^* = 0.8465$ is obtained on $\mathcal{A}_3$ with $R_3^{\text{eoi}} = 0.9946$. It shows that the feedback of $\mathcal{A}_1$ at the 7799th step is expectedly better than $\mathcal{A}_3$'s at the 9946th step. Focusing on maximizing the final feedback, $\mathcal{A}_1$ is the best.
- $\mathcal{A}_3$ easily misleads the compared methods that don't consider maximizing the long-term feedback. C-UCB, Softmax, ε-greedy, and UCB-E allocate the majority of trials to $\mathcal{A}_3$.
- In the non-stationary setting, the methods that consider maximizing extreme values tend to explore more. The R_i^{eoi} in Extreme and ER-UCB-S are uniform across arms.
- The extreme region target is important in this setting. ER-UCB-N doesn't select $\mathcal{A}_2$ but accurately selects $\mathcal{A}_1$ and allocates the majority of trials to $\mathcal{A}_1$.

13.3.2 AutoML Tasks

The AutoML tasks consist of datasets and a set of learning algorithms. Table 13.3 shows the details of the selected learning algorithms. We select a total of 6 datasets from UCI. The evaluation criterion is the fivefold cross-validation accuracy. To construct the stationary setting of AutoML, we select random search as the solver for hyper-parameter optimization. To construct the non-stationary setting of AutoML, we use SRACOS as the solver for hyper-parameter optimization. Besides the compared methods, we apply random search and SRACOS to search on the joint hyper-parameter space as the baseline (Joint). We set the total number of trials to $n = 200$. All experiments are run for 15 independent times. We report the average performance. When

Table 13.3 Details of the hyper-parameters of the candidate classification algorithms [11]. #Int. is the number of integer hyper-parameters. #Cont. is the number of continuous hyper-parameters. #Cate. is the number of categorical hyper-parameters. #$\boldsymbol{\Delta}$ is the dimensionality of the whole hyper-parameter space

Algorithms	#Int.	#Cont.	#Cate.	#$\boldsymbol{\Delta}$
DecisionTree (DT)	3	0	2	5
ExtraTree (ET)	3	0	2	5
Kneighbors (KN)	1	0	2	3
PassiveAggressive (PA)	0	2	3	5
AdaBoost (Ada)	1	1	1	3
Bagging (Bag)	1	0	0	1
RandomForest (RF)	2	1	3	6
GaussianNB (NB)	0	1	0	1

the solver is random search, we set $\phi(x) = x$ for ER-UCB-N. When the solver is SRacos, we set $\phi(x) = 2/(1 + \exp(-0.005x)) - 1$ for ER-UCB-N. For both task settings, we set $\theta = 0.01$, $\gamma = 20$, and $\beta = 0.6$ for ER-UCB-S, and $\theta = 0.01$ and $\gamma = 20$ for ER-UCB-N.

According to Table 13.4, we can draw the following conclusions:

- The "no free lunch" theorem has been proved again. There is no algorithm that can beat others on all 6 datasets. Thus, we have to select a suitable algorithm for each dataset. In other words, algorithm selection is necessary for the AutoML process. However, ensemble classification algorithms such as AdaBoost, Bagging, RandomForest, etc., show great power in the experiments. They win the best algorithm on 4 out of 6 datasets, indicating that ensemble algorithms usually have good robustness.
- DFO shows better efficiency than random search for hyper-parameter optimization. DFO beats random search on 4/6 datasets, indicating that employing DFO methods to tune hyper-parameters is more reasonable.
- The cascaded algorithm selection framework can effectively improve the performance of the AutoML process. The methods based on the cascaded algorithm selection framework usually receive better validation accuracy than the methods based on joint search space because the cascaded algorithm selection framework successfully avoids the redundant dimension issue and improves the efficiency of hyper-parameter tuning.
- On stationary arms, ER-UCB-S has the best performance. The results of stationary arms are located on the left part of Table 13.4 using random search. From the column of $\bar{X}^*$, ER-UCB-S outperforms others on 6/6 datasets. From the column of X^*, ER-UCB-S wins on 5/6 datasets. Furthermore, ER-UCB-S obtains $\mathcal{A}_{X^*} = \tilde{\mathcal{A}}_i$ on 6/6 datasets, indicating that ER-UCB-S can precisely find the true best algorithm and allocate trials to it. Compared with other methods, the extreme region target effectively helps ER-UCB-S find algorithms that can potentially reach extreme values.
- On non-stationary arms, ER-UCB-N has the best performance. On non-stationary arms (right part of Table 13.4, using DFO), ER-UCB-N outperforms others on 6/6 datasets. And ER-UCB-N obtains $\mathcal{A}_{X^*} = \tilde{\mathcal{A}}_i$ on 6/6 datasets. Compared with ER-UCB-S, ER-UCB-N has better stability, indicating that the convergence curve estimation of ER-UCB-N effectively leads ER-UCB-N to correct arm selection.
- ER-UCB-N also shows competitive power on stationary arms. By setting a linear function $\phi(x) = x$, ER-UCB-N receives the best performance (X^*) on 1/6 datasets. On other datasets, ER-UCB-N also outperforms most of the compared methods. But the $R^{\mathrm{eoi}}_{\tilde{\mathcal{A}}_i}$ of ER-UCB-N stays at a low level. ER-UCB-N needs some trials to estimate the convergence curve. But the curve estimation is not necessary on stationary arms. The exploration action wastes some trials and has a negative effect on trial allocation.

Table 13.4 Performances on 6 AutoML tasks [11]. $\bar{X}^*$ is the average performance of the three repeated runs, among which, X^* is the best one. $\mathcal{A}_{X^*}$ is the selected algorithm. $\tilde{\mathcal{A}}_i = \arg\max_{\mathcal{A}_i \in \boldsymbol{\mathcal{A}}} R^{\text{eoi}}_{\mathcal{A}_i}$ denotes the algorithm that receives the majority of the trials. $R^{\text{eoi}}_{\tilde{\mathcal{A}}_i}$ is the average exploitation rate on $\tilde{\mathcal{A}}_i$. In the columns of $\bar{X}^*$ and X^*, the number in bold is the best performance among the compared methods

Dataset	Random search						Derivative-free optimization					
	Method	$\bar{X}^*$	X^*	$\mathcal{A}_{X^*}$	$\tilde{\mathcal{A}}_i$	$R^{\text{eoi}}_{\tilde{\mathcal{A}}_i}$	Method	$\bar{X}^*$	X^*	$\mathcal{A}_{X^*}$	$\tilde{\mathcal{A}}_i$	$R^{\text{eoi}}_{\tilde{\mathcal{A}}_i}$
Adult	ER-UCB-N	0.8714 ± 0.0003	0.8717	Ada	Ada	0.32 ± 0.15	ER-UCB-N	**0.8732** ± 0.0005	**0.8738**	Ada	Ada	0.58 ± 0.05
	ER-UCB-S	**0.8720** ± 0.0004	**0.8723**	Ada	Ada	0.45 ± 0.01	ER-UCB-S	0.8724 ± 0.0007	0.8732	Ada	Ada	0.44 ± 0.23
	Extreme	0.8716 ± 0.0001	0.8717	Ada	Ada	0.38 ± 0.00	Extreme	0.8722 ± 0.0003	0.8724	Ada	Ada	0.25 ± 0.00
	C-UCB	0.8715 ± 0.0005	0.8719	Ada	Ada	0.32 ± 0.03	C-UCB	0.8715 ± 0.0002	0.8717	Ada	Ada	0.23 ± 0.01
	ε-greedy	0.8696 ± 0.0028	0.8722	Ada	RF	0.65 ± 0.17	ε-greedy	0.8726 ± 0.0003	0.8728	Ada	Ada	0.35 ± 0.03
	Softmax	0.8711±0.0003	0.8715	Ada	Bag	0.13 ± 0.08	Softmax	0.8718 ± 0.0001	0.8719	Ada	Ada	0.17 ± 0.07
	S-Halving	0.8713 ± 0.0003	0.8718	Ada	Ada	0.29 ± 0.00	S-Halving	0.8700 ± 0.0014	0.8719	Ada	Ada	0.29 ± 0.00
	UCB-E	0.8711 ± 0.0005	0.8718	Ada	Bag	0.17 ± 0.01	UCB-E	0.8698 ± 0.0016	0.8717	Ada	RF	0.18 ± 0.01
	Random	0.8707 ± 0.0002	0.8710	Ada	RF	0.15 ± 0.01	Random	0.8714 ± 0.0008	0.8722	Ada	RF	0.15 ± 0.01
	Joint	0.8698 ± 0.0002	0.8699	Ada	–	–	Joint	0.8719 ± 0.0003	0.8725	Ada	–	–
	AutoSKL	–	–	–	–	–	AutoSKL	0.8715 ± 0.0003	0.8719	Ada	–	–
Balance	ER-UCB-N	0.8902 ± 0.0042	0.8952	PA	PA	0.50 ± 0.34	ER-UCB-N	**0.8920** ± 0.0032	**0.8958**	PA	PA	0.88 ± 0.09
	ER-UCB-S	**0.8933** ± 0.0038	**0.8981**	PA	PA	0.92 ± 0.01	ER-UCB-S	0.8828 ± 0.0068	0.8878	PA	PA	0.52 ± 0.28
	Extreme	0.8797 ± 0.0000	0.8797	KN	KN	0.65 ± 0.00	Extreme	0.8812 ± 0.0006	0.8816	KN	KN	0.65 ± 0.00
	C-UCB	0.8886 ± 0.0034	0.8918	PA	PA	0.84 ± 0.04	C-UCB	0.8886 ± 0.0042	0.8937	PA	PA	0.58 ± 0.10
	ε-greedy	0.8874 ± 0.0052	0.8928	PA	PA	0.76 ± 0.13	ε-greedy	0.8918 ± 0.0036	**0.8958**	PA	PA	0.88 ± 0.01
	Softmax	0.8786 ± 0.0042	0.8857	PA	KN	0.14 ± 0.01	Softmax	0.8798 ± 0.0024	0.8837	PA	GNB	0.16 ± 0.04
	S-Halving	0.8846 ± 0.0048	0.8897	PA	PA	0.29 ± 0.00	S-Halving	0.8881 ± 0.0049	0.8938	PA	PA	0.29 ± 0.00
	UCB-E	0.8827 ± 0.0051	0.8918	PA	PA	0.30 ± 0.01	UCB-E	0.8862 ± 0.0058	0.8918	PA	PA	0.36 ± 0.01
	Random	0.8768 ± 0.0084	0.8840	PA	KN	0.15 ± 0.02	Random	0.8832 ± 0.0046	0.8878	PA	Ada	0.13 ± 0.01
	Joint	0.8890 ± 0.0040	0.8918	PA	–	–	Joint	0.8886 ± 0.0040	0.8918	PA	–	–
	AutoSKL	–	–	–	–	–	AutoSKL	0.8777 ± 0.0033	0.8816	KN	–	–

(continued)

Table 13.4 (continued)

Dataset	Random search						Derivative-free optimization					
	Method	$\bar{X}^*$	X^*	$\mathcal{A}_{X^*}$	$\tilde{\mathcal{A}}_i$	$R^{\text{eoi}}_{\tilde{\mathcal{A}}_i}$	Method	$\bar{X}^*$	X^*	$\mathcal{A}_{X^*}$	$\tilde{\mathcal{A}}_i$	$R^{\text{eoi}}_{\tilde{\mathcal{A}}_i}$
Car	ER-UCB-N	0.8564 ± 0.0046	0.8628	Bag	Ada	0.17 ± 0.01	ER-UCB-N	**0.8692** ± 0.0042	**0.8748**	Bag	Bag	0.90 ± 0.06
	ER-UCB-S	**0.8596** ± 0.0086	**0.8683**	Bag	Bag	0.93 ± 0.01	ER-UCB-S	0.8680 ± 0.0048	0.8733	DT	Ada	0.34 ± 0.40
	Extreme	0.8558 ± 0.0082	0.8648	Bag	Bag	0.16 ± 0.00	Extreme	0.8648 ± 0.0072	0.8703	ET	DT	0.36 ± 0.39
	C-UCB	0.8578 ± 0.0078	0.8664	Bag	Bag	0.87 ± 0.02	C-UCB	0.8583 ± 0.0048	0.8642	Bag	Bag	0.43 ± 0.23
	ε-greedy	0.8574 ± 0.0068	0.8642	Bag	Bag	0.87 ± 0.01	ε-greedy	0.8572 ± 0.0056	0.8639	Bag	Bag	0.37 ± 0.30
	Softmax	0.8568 ± 0.0048	0.8625	Ada	Bag	0.13 ± 0.01	Softmax	0.8563 ± 0.0051	0.8625	Ada	Bag	0.15 ± 0.02
	S-Halving	0.8562 ± 0.0048	0.8639	Bag	Bag	0.29 ± 0.00	S-Halving	0.8553 ± 0.0047	0.8676	Bag	Bag	0.29 ± 0.00
	UCB-E	0.8572 ± 0.0058	0.8625	Bag	Bag	0.33 ± 0.01	UCB-E	0.8559 ± 0.0043	0.8646	Bag	Bag	0.39 ± 0.01
	Random	0.8556 ± 0.0042	0.8618	Bag	Ada	0.15 ± 0.02	Random	0.8636 ± 0.0026	0.8654	DT	GNB	0.14 ± 0.01
	Joint	0.8562 ± 0.0070	0.8632	DT	–	–	Joint	0.8617 ± 0.0089	0.8719	Bag	–	–
	AutoSKL	–	–	–	–	–	AutoSKL	0.8614 ± 0.0055	0.8675	ET	–	–
Chess	ER-UCB-N	0.9457 ± 0.0086	0.9562	Ada	Ada	0.48 ± 0.23	ER-UCB-N	**0.9524** ± 0.0228	**0.9761**	Ada	Ada	0.70 ± 0.03
	ER-UCB-S	**0.9498** ± 0.0118	**0.9621**	Ada	Ada	0.57 ± 0.12	ER-UCB-S	0.9358 ± 0.0074	0.9434	Ada	Ada	0.30 ± 0.32
	Extreme	0.9352 ± 0.0068	0.9405	DT	DT	0.96 ± 0.00	Extreme	0.9364 ± 0.0065	0.9405	DT	DT	0.96 ± 0.00
	C-UCB	0.9383 ± 0.0022	0.9405	PA	PA	0.40 ± 0.01	C-UCB	0.9352 ± 0.0072	0.9428	Ada	Ada	0.40 ± 0.08
	ε-greedy	0.9392 ± 0.0021	0.9413	PA	Ada	0.55 ± 0.18	ε-greedy	0.9408 ± 0.0024	0.9436	Ada	Ada	0.49 ± 0.28
	Softmax	0.9348 ± 0.0028	0.9386	PA	PA	0.15 ± 0.10	Softmax	0.9413 ± 0.0039	0.9456	Ada	RF	0.15 ± 0.04
	S-Halving	0.9383 ± 0.0032	0.9428	PA	PA	0.29 ± 0.00	S-Halving	0.9408 ± 0.0084	0.9436	PA	PA	0.29 ± 0.00
	UCB-E	0.9368 ± 0.0053	0.9416	PA	Ada	0.21 ± 0.01	UCB-E	0.9412 ± 0.0098	0.9508	Ada	PA	0.23 ± 0.01
	Random	0.9336 ± 0.0052	0.9396	Ada	PA	0.14 ± 0.01	Random	0.9375 ± 0.0048	0.9413	Ada	KN	0.14 ± 0.01
	Joint	0.9423 ± 0.0050	0.9436	Ada	–	–	Joint	0.9421 ± 0.0070	0.9448	PA	–	–
	AutoSKL	–	–	–	–	–	AutoSKL	0.9392 ± 0.0008	0.9413	PA	–	–

(continued)

Table 13.4 (continued)

Dataset	Random search						Derivative-free optimization					
	Method	$\bar{X}^*$	X^*	$\mathcal{A}_{X^*}$	$\tilde{\mathcal{A}}_i$	$R^{\text{eoi}}_{\tilde{\mathcal{A}}_i}$	Method	$\bar{X}^*$	X^*	$\mathcal{A}_{X^*}$	$\tilde{\mathcal{A}}_i$	$R^{\text{eoi}}_{\tilde{\mathcal{A}}_i}$
Credit	ER-UCB-N	0.9124 ± 0.0048	0.9164	DT	DT	0.43 ± 0.32	ER-UCB-N	**0.9144** ± 0.0036	**0.9182**	ET	ET	0.54 ± 0.32
	ER-UCB-S	**0.9152** ± 0.0026	**0.9182**	RF	RF	0.53 ± 0.03	ER-UCB-S	0.9128 ± 0.0000	0.9128	Ada	RF	0.50 ± 0.06
	Extreme	0.9128 ± 0.0000	0.9128	DT	DT	0.96 ± 0.00	Extreme	0.9042 ± 0.0078	0.9128	DT	DT	0.96 ± 0.00
	C-UCB	0.9128 ± 0.0000	0.9128	RF	RF	0.44 ± 0.01	C-UCB	0.9128 ± 0.0000	0.9128	DT	RF	0.57 ± 0.09
	ε-greedy	0.9108 ± 0.0012	0.9128	RF	RF	0.58 ± 0.36	ε-greedy	0.9086 ± 0.0035	0.9128	RF	RF	0.49 ± 0.38
	Softmax	0.9086 ± 0.0042	0.9128	ET	RF	0.14 ± 0.02	Softmax	0.9128 ± 0.0000	0.9128	DT	DT	0.15 ± 0.01
	S-Halving	0.9117 ± 0.0039	0.9128	ET	ET	0.29 ± 0.00	S-Halving	0.9115 ± 0.0027	0.9146	ET	ET	0.29 ± 0.00
	UCB-E	0.9122 ± 0.0023	0.9145	ET	Bag	0.20 ± 0.01	UCB-E	0.9101 ± 0.0035	0.9128	RF	Bag	0.22 ± 0.01
	Random	0.9128 ± 0.0000	0.9128	ET	Ada	0.15 ± 0.01	Random	0.9128 ± 0.0000	0.9128	Ada	Bag	0.14 ± 0.01
	Joint	0.9128 ± 0.0000	0.9128	ET	–	–	Joint	0.9136 ± 0.0024	0.9164	ET	–	–
	AutoSKL	–	–	–	–	–	AutoSKL	0.9128 ± 0.0000	0.9128	DT	–	–
Spam	ER-UCB-N	0.9316 ± 0.0043	**0.9358**	RF	RF	0.47 ± 0.33	ER-UCB-N	**0.9363** ± 0.0033	**0.9401**	RF	RF	0.85 ± 0.11
	ER-UCB-S	**0.9322** ± 0.0008	0.9331	RF	RF	0.38 ± 0.11	ER-UCB-S	0.9338 ± 0.0043	0.9377	RF	RF	0.53 ± 0.09
	Extreme	0.9201 ± 0.0048	0.9271	Ada	DT	0.96 ± 0.00	Extreme	0.9212 ± 0.0026	0.9285	Bag	DT	0.96 ± 0.00
	C-UCB	0.9289 ± 0.0008	0.9302	Ada	Ada	0.40 ± 0.08	C-UCB	0.9302 ± 0.0002	0.9304	Ada	Ada	0.46 ± 0.04
	ε-greedy	0.9292 ± 0.0012	0.9314	Ada	Ada	0.48 ± 0.40	ε-greedy	0.9293 ± 0.0019	0.9320	Ada	Bag	0.59 ± 0.41
	Softmax	0.9293 ± 0.0005	0.9297	RF	RF	0.14 ± 0.01	Softmax	0.9303 ± 0.0024	0.9339	Ada	RF	0.14 ± 0.01
	S-Halving	0.9292 ± 0.0014	0.9304	RF	RF	0.29 ± 0.00	S-Halving	0.9283 ± 0.0017	0.9331	RF	RF	0.29 ± 0.00
	UCB-E	0.9292 ± 0.0013	0.9309	RF	Ada	0.20 ± 0.01	UCB-E	0.9276 ± 0.0028	0.9323	Ada	Ada	0.24 ± 0.01
	Random	0.9287 ± 0.0003	0.9292	Ada	RF	0.15 ± 0.01	Random	0.9298 ± 0.0014	0.9317	RF	RF	0.14 ± 0.01
	Joint	0.9258 ± 0.0011	0.9288	Ada	–	–	Joint	0.9310 ± 0.0024	0.9358	RF	–	–
	AutoSKL	–	–	–	–	–	AutoSKL	0.9320 ± 0.0022	0.9358	RF	–	–

13.4 Summary

For algorithm selection tasks, we present the cascaded algorithm selection framework in this chapter to tackle the redundancy issue of the joint hyper-parameter space. Cascaded algorithm selection has a two-level process. The lower level is the hyper-parameter optimization process that searches hyper-parameters in the space of a learning algorithm but not the joint space of all algorithms. The upper level is formulated as a multi-armed bandit task, in which a hyper-parameter optimization process of a learning algorithm can be seen as an arm. The key to the bandit is a strategy that finds the best arm and allocates trials to it as many as possible. AutoML needs to find the best algorithm and its best hyper-parameter configuration.

With this target in mind, this chapter presents the Extreme-Region Upper-Confidence Bound (ER-UCB) strategy to maximize the final feedback by selecting the arm with the largest extreme region. We design ER-UCB-S and ER-UCB-N algorithms for stationary and non-stationary feedback distributions. With K arms and n total trials, theoretical study shows that ER-UCB-S has an $O(K \ln n)$ upper bound and ER-UCB-N has an $O(Kn^{\nu})$ upper bound on the extreme region regret, where $\frac{2}{3} < \nu < 1$. We also investigated the ER-UCB strategy on synthetic and AutoML tasks. The empirical results verified the effectiveness of the presented methods.

References

1. Adankon MM, Cheriet M (2009) Model selection for the LS-SVM. application to handwriting recognition. Pattern Recognit **42**(12), 3264–3270 (2009)
2. Bengio Y (2000) Gradient-based optimization of hyperparameters. Neural Comput 12(8):1889–1900
3. Bergstra J, Bengio Y (2012) Random search for hyper-parameter optimization. J Mach Learn Res 13:281–305
4. Biem A (2003) A model selection criterion for classification: application to hmm topology optimization. In: Proceedings of the 7th international conference on document analysis and recognition, pp 104–108
5. Brazdil PB, Soares C, Da Costa JP (2003) Ranking learning algorithms: Using IBL and meta-learning on accuracy and time results. Mach Learn 50(3):251–277
6. Bubeck S, Cesa-Bianchi N et al (2012) Regret analysis of stochastic and nonstochastic multi-armed bandit problems. Found Trends Mach Learn 5(1):1–122
7. Carpentier A, Valko M (2014) Extreme bandits. In: Advances In neural information processing systems, pp 1089–1097
8. Guo X, Yang J, Wu C, Wang C, Liang Y (2008) A novel LS-SVMs hyper-parameter selection based on particle swarm optimization. Neurocomputing 71(16):3211–3215
9. Hu Y-Q, Yu Y, Zhou Z-H (2018) Experienced optimization with reusable directional model for hyper-parameter search. In: Proceeding of the 27th international joint conference on artificial intelligence, pp 2276–2282
10. Hu YQ, Liu Z, Yang H, Yu Y, Liu Y (2020) Derivative-free optimization with adaptive experience for efficient hyper-parameter tuning. In: Proceeding of the 24th European conference on artificial intelligence, pp 1207–1214
11. Hu Y-Q, Liu X-H, Li S-Q, Yu Y (2022) Cascaded algorithm selection with extreme-region ucb bandit. IEEE Trans Pattern Anal Mach Intell 44(10):6782–6794

12. Hutter F, Hoos HH, Leyton-Brown K (2011) Sequential model-based optimization for general algorithm configuration. LION 5:507–523
13. Shahriari B, Swersky K, Wang Z, Adams RP, Freitas ND (2015) Taking the human out of the loop: a review of Bayesian optimization. Proc IEEE 104(1):148–175
14. Snoek J, Larochelle H, Adams RP (2012) Practical Bayesian optimization of machine learning algorithms. In: Advances in neural information processing systems, vol 25, Lake Tahoe, Nevad, pp 2960–2968a
15. Sutton RS, Barto AG (1998) Reinforcement learning: an introduction. MIT Press, Cambridge, MA
16. Tokic M, Palm G (2011) Value-difference based exploration: adaptive control between epsilon-greedy and softmax. In: Annual conference on artificial intelligence. Springer, pp 335–346
17. Yao Q, Wang M, Chen Y, Dai W, Li YF, Tu WW, Yang Q, Yu Y (2018) Taking human out of learning applications: A survey on automated machine learning. arXiv:1810.13306

Chapter 14
Calculation Operation Optimization: Competition Neural Architecture Search

Abstract This chapter introduces Competition Neural Architecture Search (CNAS), a method for automatically designing neural network architectures. CNAS separates the search process into two parts: topological structure enumeration and calculation operation optimization. The topological structures are enumerated under depth and width constraints, while the calculation operations are optimized using derivative-free optimization (DFO) methods. A competition mechanism is employed to iteratively eliminate poorly performing structures, ensuring that the best architecture is selected. To improve efficiency, CNAS uses block-based search and experience reuse, leveraging historical data to warm-start the optimization process and simulate competitions. The chapter presents empirical results on image classification and denoising tasks, demonstrating that CNAS achieves competitive performance compared to manual designs and state-of-the-art NAS methods. The experiments highlight CNAS's ability to efficiently explore the architecture space and produce high-quality network designs.

Neural Architecture Search (NAS) [11, 20] has been proposed to automatically design neural network architectures for deep learning tasks. This process can be seen as determining a topological structure and operation settings on this structure. Recent NAS methods consider both parts simultaneously, which is usually hard to thoroughly explore the architecture space. This chapter presents a competition neural architecture search framework that considers topological structure search and operation setting search separately. The method enumerates all possible topological structures within limited length and width settings. For each structure, derivative-free optimization, introduced in the previous chapters, is utilized to optimize its operation setting. A competition mechanism is proposed to combine these two parts. In each optimization loop, the topological structures are compared with each other according to their previous operation setting performance, and the bad structures are eliminated. The structure that is left finally with its best operation setting is the search result. This chapter also presents an experience-reused mechanism to accelerate the search process. With manual architectures and historical architectures, experience can be extracted to preliminarily screen structures and warmly start the operation search

Y. Yu et al., *Derivative-Free Optimization*, Machine Learning: Foundations, Methodologies, and Applications, https://doi.org/10.1007/978-981-96-5929-6_14

processes. The experiments on image classification tasks and an image denoising task show that the presented method receives competitive architectures compared with some manual networks and state-of-the-art NAS methods.

14.1 Calculation Operation Optimization with Neural Architecture Search

NAS has developed along with the whole development of neural networks. In the 1990s, researchers tried to apply evolutionary algorithms to design the connections for fully-connected networks and tune the hyper-parameters of their training processes [18]. Because fully-connected networks can be easily formulated by some hyper-parameters such as connection types, layer size, neural size, etc., the architecture design is considered a hyper-parameter optimization problem [2, 3]. Evolutionary algorithms are suitable as the solver for this problem. However, many irregular and imaginative architectures have been invented for different learning tasks. This causes the architecture search space to become more complex and hard to be formulated by several simple hyper-parameters. Thus, recent NAS works make more contributions to the architecture space design and efficient search method proposal.

Previous NAS methods [10, 19, 20] usually consider a Directed Acyclic Graph (DAG) space for the architecture search. A DAG consists of edges and nodes. The edges indicate the data flows in network architectures, while the nodes indicate the calculation operations. With a limited setting of depth and width, the DAG space contains any possible network architectures. Because the number of candidate architectures expands exponentially with the increase of depth and width, previous works place restrictions on the search space in two ways: macro space and micro space. The macro space [1, 20] aims to generate entire network architectures directly and intuitively. Thus, this search space always has a large depth setting. To limit the size of the search space, the macro space only allows single-chain styled architectures. Based on this, skipping connections are also allowed to generate complex structures. However, the macro space has clear shortcomings. It is impossible to generate deep network architectures, and the search space is still huge with a large setting of depth. Thus, the micro space has been proposed in NASNet [21]. The micro space aims at a part of network architectures, i.e., a block, but not entire architectures, and then stacks them together to construct an entire architecture. For example, NASNet only searches for a normal block and a reduction block and repeats them 5 times to construct a network architecture. It is a narrow search space for a block search. Furthermore, the operation of stacking blocks effectively generates deep network architectures. Thus, more and more NAS works [9–12, 16] apply the micro space to search for effective network architectures.

14.1.1 NAS Task Formulation

In this chapter, we only solve the architecture search on CNN. A deep learning task can be defined by a training dataset $\mathcal{D}_{\text{train}}$ and a testing dataset $\mathcal{D}_{\text{test}}$. Let $(\boldsymbol{x}, \boldsymbol{y}) \in \mathcal{D}_{\text{train}} \cup \mathcal{D}_{\text{test}}$ denote an instance, where $\boldsymbol{x}$ is an input tensor and $\boldsymbol{y}$ is its corresponding label tensor. A network architecture can be seen as a DAG that consists of edges and nodes. The edges are the data flows in the network architecture, indicating the topological structure. The nodes are the transformations from the input data flows to output data flows, indicating the calculation operations. We assume a DAG has m nodes, and let $F = \{f_1, f_2, \ldots, f_m\}$ denote a set of calculation operations. Any $f \in F$ is a calculation operation that transforms the input tensors to an output tensor, i.e., $\boldsymbol{x}^{\text{out}} = f(\boldsymbol{x}_1^{\text{in}}, \ldots)$. We note that the number of input tensors may be more than one, but the number of output tensors must be one. By setting a nested connection, we can determine a topological structure. Let $f^{\text{out}} \otimes_i (f_1^{\text{in}}, \ldots)$ denote the nested connection in the ith node. Thus, $\boldsymbol{\otimes} = \{\otimes_1, \otimes_2, \ldots, \otimes_m\}$ denotes a topological structure. A network architecture can be presented as follows:

$$\mathcal{N} = \boldsymbol{\otimes} \lhd F, \tag{14.1}$$

where $\boldsymbol{\otimes}$ is a topological structure and F is a set of calculation operations, and the symbol $\lhd$ means correlating F to $\boldsymbol{\otimes}$. Let $\boldsymbol{\mathcal{N}} = \{\mathcal{N}_1, \mathcal{N}_2, \ldots\}$ denote the set of all possible network architectures. The NAS task can be formulated as follows:

$$\mathcal{N}^* = \underset{\mathcal{N} \in \boldsymbol{\mathcal{N}}}{\arg\max}\, C_{\text{valid}}(\mathcal{N}, \mathcal{D}_{\text{train}}), \tag{14.2}$$

where $C_{\text{valid}}(\cdot, \mathcal{D}_{\text{train}})$ is an evaluation criterion of the validation process on the training dataset. The target of NAS is to select the network architecture from the possible network architecture set that has the maximal validation performance.

We present the method of *Competition Neural Architecture Search* (CNAS) with reused experience. CNAS is a hierarchical search process, i.e., considering the topological structure enumeration and calculation operation optimization separately. And a competition mechanism is applied to combine both of them.

14.1.2 Topological Structure Enumeration

With limited settings of depth (α) and width (β), we can enumerate all possible topological structures. To avoid obtaining strange structures, we apply a main-and-branch chain approach for enumeration. In this approach, we set a main chain first, which is the deepest chain in the topological structure. The length of the main chain corresponds to the depth setting. Then, we add branch chains to the main chain. The number of branch chains corresponds to the width setting. There are some constraints when adding branches. First, skipping connections are not considered

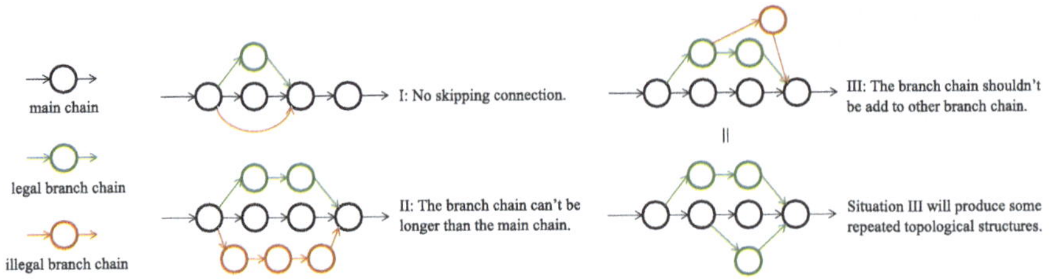

Fig. 14.1 Illustration of the constraints when enumerating topological structures. The circle and arrow in black indicate the main chain. The circle and arrow in green indicate the legal branch chain. The circle and arrow in red indicate the illegal branch chain

when enumerating. Under this constraint, the branch chain must have at least one node. Second, the length of the branch chain can't be larger than the main chain's. Otherwise, it will produce a branch chain that is longer than the main chain. Third, it is forbidden to add a branch chain to another branch chain because some repeated and uncontrollable structures will be constructed in this way. Figure 14.1 shows the constraints when enumerating topological structures. Under these constraints, we can enumerate possible topological structures that are non-repeated and reasonable. We denote the enumeration process as $\bigotimes = \{\otimes_1, \otimes_2, \ldots\} = \texttt{enumerate}(\alpha, \beta)$, where $\otimes_i \in \bigotimes$ is a possible topological architecture.

Although we constrain the enumeration process, the number of possible topological structures increases explosively when depth and width become large. To tackle this, a block-based search is employed, which we will discuss later.

14.1.3 Calculation Operation Optimization

The enumeration process gives the topological structures. The next process should assemble the calculation operations for each structure. We consider it a black-box optimization task. A node in the structure is a calculation operation that transforms tensors from input to output. In CNN, the options for calculation operations are limited and easy to parameterize. For example, we show a parameterized search space of calculation operation optimization in Table 14.1. t_C is the type of calculation operation. We only consider convolutional or pooling operations. f is the kernel size. We set some candidates for it, which are 2×2, 3×3, or 5×5. d is the filter size. a is the activation type. t_P is the pooling type. Average pooling or maximal pooling are considered. s is the skipping connection setting. We design the skipping connection architecture by this parameter. The parameter value means how many nodes will be skipped to input the output of this node. $\emptyset$ means no skipping connection. Figure 14.2 gives an example of the correspondence between parameterized code and network architecture. We note that not all dimensions of a setting are active. When $t_C =$ Conv., dimensions 1, 2, 3, 4, and 6 are active. When $t_C =$ Pool., dimensions 1,

Table 14.1 The search space settings of calculation operations

Dim.	Symbol	Parameter	Setting
1	t_C	Type	{Conv., Pool.}
2	f	Kernel size	Square. $\in \{2, 3, 5\}$
3	d	Filter size	$\in \{16, 24, 32\}$
4	a	Activation	{ReLU, LeakyReLU, ReLU6}
5	t_P	Pooling	{ave., max.}
6	s	Skipping	$\{\emptyset, 1, 2, 3\}$

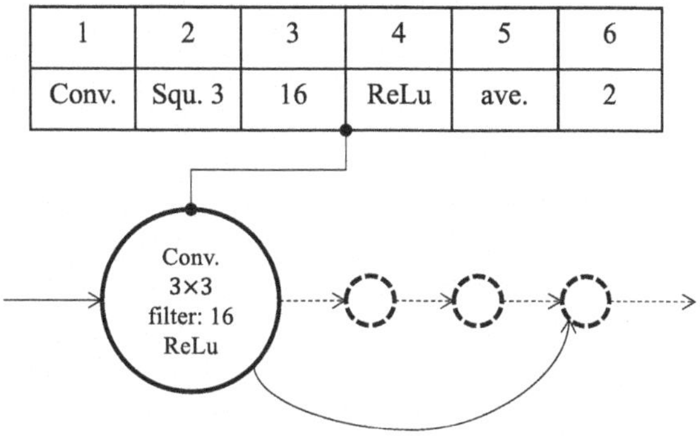

Fig. 14.2 Illustration of an example of a calculation operation setting. The table shows the code in the parameterized search space. The figure shows the corresponding network architecture

2, and 5 are active. In this way, one node can be parameterized by 6 dimensions. Let N denote the number of nodes of a $\otimes$ and $\mathcal{F}_\otimes$ denote the calculation operation space. Thus, $|\mathcal{F}_\otimes| = 6 \times N$. Because $\otimes$ won't have too many nodes, the optimization won't suffer from a high-dimensional issue.

Like hyper-parameter tuning, calculation operation optimization can be seen as a black-box optimization problem and solved by DFO methods, introduced in the previous chapters. DFO relies only on the evaluation values, not the gradients, to explore the search space. Thus, it is more suitable for this problem. Currently, popular DFO methods are evolutionary algorithms [4], Bayesian optimization [8, 14], classification-based optimization [7], etc. These methods all follow a sample-evaluation framework. Let $\omega_\otimes$ denote a calculation operation optimization process for $\otimes$. $\omega_\otimes$ has an inner model that is used to sample a new configuration of calculation operations: $F = \omega_\otimes.\texttt{sample}(\mathcal{F}_\otimes)$. An entire network architecture can be constructed: $\mathcal{N} = \otimes \lhd F$. Now, a criterion is needed to evaluate the quality of this network architecture. The evaluation is usually a validation process as follows:

$$
\begin{aligned}
v = C_{\text{valid}}(\mathcal{N}^{\boldsymbol{w}^*}, \mathcal{D}_{\text{train}}), \\
\text{s.t. } \boldsymbol{w}^* = \arg\min_{\boldsymbol{w}} \mathcal{L}_{\text{train}}(\mathcal{N}^{\boldsymbol{w}}, \mathcal{D}_{\text{train}}),
\end{aligned} \tag{14.3}
$$

where C_{valid} is the validation criterion, such as accuracy, F1 score, etc., and $\mathcal{L}_{\text{train}}$ is the training loss, such as mean square loss, cross-entropy, etc. With the evaluation value v, we update the inner model and get ready for the next sampling: $\omega_{\otimes} = \omega_{\otimes}.\texttt{update}(F, v)$. It is a stepwise framework that can be paused at any optimization loop. And the optimization process can be activated if necessary. This mechanism makes it possible to design a competition approach to combine the calculation operation optimization and topological structure enumeration.

14.2 The CNAS Algorithm

In this step, we have to combine the topological structure enumeration and calculation operation optimization together. The target is to eliminate the topological structures that have poor performance during optimizing their calculation operations. A competition mechanism is employed to accomplish this task. For each topological structure $\otimes_i$, there is an optimization process $\omega_{\otimes_i}$ that searches for its best calculation operations. The optimization process includes a sample step, an evaluation step, and an update step. This optimization process can be paused at any loop, as we mentioned before. Let $\tilde{F}_i$ denote the best-so-far calculation operation setting of $\otimes_i$. After B loops for each optimization process, the 1st round competition can be raised by comparing $C_{\text{valid}}(\otimes_i \lhd \tilde{F}_i, \mathcal{D}_{\text{train}})$ with each other. And half of the topological structures that lose the competition will be eliminated. After K rounds, there is only one topological structure left, which is the winner topological structure. Then, we employ more loops (B') to intensively optimize its calculation operations. This process is shown in Fig. 14.3. The combination of the final best-so-far calculation operations and the winner topological structure is the winner network architecture.

There are some discussions about the competition mechanism. The competition mechanism can select the best topological structure with an exponential decay rate

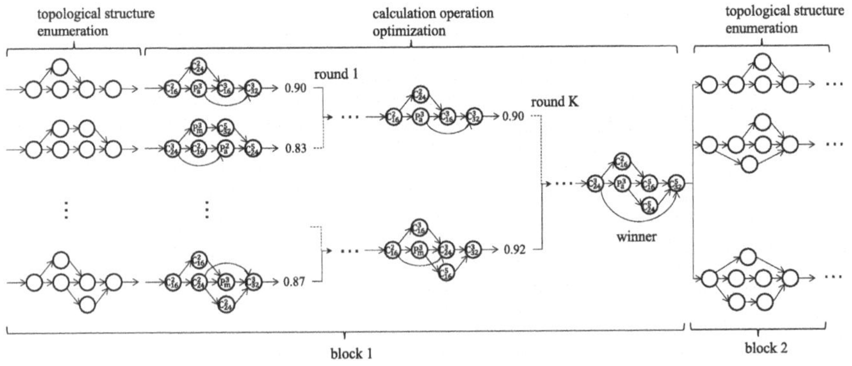

Fig. 14.3 Illustration of the framework of CNAS that includes topological structure enumeration, calculation operation optimization, competition mechanism, and block-based search. C_d^f means a convolutional operation with an $f \times f$ kernel and d filters. $P_{a/m}^f$ means an average or maximal pooling operation with an $f \times f$ kernel

because half of the structures will be eliminated in each round. It can alleviate the issue of the explosive number of possible topological structures in the enumeration phase. The competition mechanism is a greedy strategy for selecting structures. But the best structure will be selected finally. The competition process can be controlled by some hyper-parameters such as the number of loops within a round B, the number of loops for optimizing the calculation operations on the winner structure B', etc. Intuitively, larger B and B' will lead to better network architectures, but it has more cost. However, the total search cost mostly depends on the number of possible topological structures. We employ block-based search to tackle this issue.

14.2.1 Block-Based Search

Due to enumeration, the number of possible topological structures increases exponentially with increasing depth and width. It is a direct way to tackle this by giving a limitation for depth and width. However, this impedes the search for deep network architectures. CNAS employs block-based search to reduce the number of enumerations and obtain deep network architectures.

The block-based search splits the entire network architecture into several parts, i.e., blocks. And then, a stepwise search is employed to search block by block. In each block, the depth and width can be set as small numbers. In our work, we usually set 4–5 nodes in depth and 2–3 branch chains in width. And we can obtain a deep enough network architecture by using only 5–6 blocks. With this setting, there are only hundreds of topological structures when enumerating, in which CNAS can quickly select the best structure with the competition mechanism. Different from the previous NAS methods with micro space, CNAS searches the next block based on the winner network architecture of the last block but does not apply repetitive blocks, as Fig. 14.3 shows. It helps us search on a larger space and obtain more suitable network architectures.

14.2.2 Experience Reuse

The search method design of CNAS considers some limitations to improve the search efficiency. But we still want to further improve the efficiency, so an experience reused approach is employed. In this work, we consider the experience from two sources. The first source is a set of manual network architectures. The manual network architectures have been proven effective by real-world applications. There are many inspired architectures that we can draw lessons from when designing the search space, such as inceptions [15], skipping connections [5, 17], linear bottlenecks [6, 13], etc. The second source is the historical log when searching network architectures. During the search, CNAS will sample many network architectures that have been evaluated. Different from manual network architectures, not all architectures from historical logs perform well. However, it won't stop us from utilizing them to avoid the new search starting from scratch. We reuse the experience in two directions. The first is the calculation operation prediction, in which the experience is used to warm-start the calculation operation optimization process. The second is the competition simulation, in which we train a simulator from experience to preliminarily screen the topological structures.

Calculation Operation Prediction

This step aims to warm-start the calculation operation optimization process. It can be formulated as a prediction problem. The input is a topological structure, and the output is the calculation operations for nodes. In this work, we focus on nodes and predict their calculation operations node by node. The manual network architectures and the historical search logs can provide the supervised information. We introduce the details from data extraction, training predictor, and prediction.

Data extraction consists of feature design and label setting. We extract features according to the topological structure. Before giving the details of feature design, we introduce a concept, i.e., collection node. If a node has more than one input or more than one output, we name this node a collection node. The collection node indicates that the network architecture isn't a simple single chain but a complex structure. We split the features into global features and local features. The global features include the number of nodes, the number of branch chains, depth, the number of collection nodes, etc., a total of 9 features that reflect the global state of the topological structure. The local features reflect the local state of a node. Thus, the local features correlate with a single node, which includes whether it is a collection node, the depth of this node, the number of branch chains that connect to this node, some information of branch chains that are related to this node, etc., a total of 16 features. Combining both of them, a total of 25 features are considered when extracting. The label of the features is the corresponding calculation operations of the experienced network architecture. The label space is the same as Table 14.1. Let $\boldsymbol{\kappa} = \{\kappa_1, \kappa_2, \ldots, \kappa_m\} = \texttt{extract}(\boldsymbol{\otimes})$ denote an extraction process from a topological structure $\boldsymbol{\otimes}$, where $\kappa_i \in \boldsymbol{\kappa}$ is a piece of features corresponding to the node $\otimes_i \in \boldsymbol{\otimes}$.

Training predictor and prediction. Because the data extracted from network architectures is time-series, we apply an LSTM model to train a predictor. Let $\boldsymbol{\phi}$ denote the predictor. For the target topological structure, we first employ the extraction process to get the features of each node. Then, the calculation operations of $\boldsymbol{\otimes}_i$ are $F_{\boldsymbol{\otimes}} = \boldsymbol{\phi}(\boldsymbol{\otimes})$.

In the calculation operation optimization process, the DFO method has an initialization step that uniformly samples on the search space. We use ϕ to predict a calculation operation configuration that replaces the uniform sample in the initialization step. The warm-start by ϕ helps the optimization process avoid starting from scratch and improves the search efficiency.

Competition Simulation

This step aims to simulate the competition between two topological structures. The key is how to evaluate the quality of a topological structure. Equation (14.3) shows the evaluation of a network architecture. But with different calculation operation configurations, a topological structure can construct different network architectures. Let v_m denote the mean of evaluation values of these network architectures, and v_s denote the standard deviation of evaluation values of these network architectures. $v_m + v_s$ is the performance upper bound that this topological structure can get with a high probability. Thus, this value can be an evaluation of a topological structure. With this evaluation, we can design the competition simulation from data extraction, training simulator, and simulation.

Data extraction. We assume the competition is between $\otimes_1$ and $\otimes_2$. The feature extraction for topological structures is the same as in the last section, that is, $\kappa_1 = \texttt{extract}(\otimes_1)$ and $\kappa_2 = \texttt{extract}(\otimes_2)$. We combine them together $[\kappa_1; \kappa_2]$ and give it a label as follows:

$$\ell([\kappa_1; \kappa_2]) = \begin{cases} +1, & v_m^{\otimes_1} + v_s^{\otimes_1} > v_m^{\otimes_2} + v_s^{\otimes_2}, \\ -1, & v_m^{\otimes_1} + v_s^{\otimes_1} \leq v_m^{\otimes_2} + v_s^{\otimes_2}. \end{cases} \tag{14.4}$$

$([\kappa_1; \kappa_2], \ell)$ is an instance for training the simulator. The label ℓ means whether $\otimes_1$ is better than $\otimes_2$. Conversely, $([\kappa_2; \kappa_1], -\ell)$ is another instance. In the search process, two instances can be extracted according to any two topological structures. From the historical search log, we can extract a training dataset.

Training simulator and simulation. It is a supervised learning task to train the simulator from a labeled dataset. This data is similar to the text classification data. Thus, an LSTM model is applied. Let ψ denote the simulator. The competition simulation can be presented as

$$\ell = \psi([\texttt{extract}(\otimes_1); \texttt{extract}(\otimes_2)]).$$

If $\ell = +1$, $\otimes_1$ wins and $\otimes_2$ is eliminated. If $\ell = -1$, $\otimes_2$ wins and $\otimes_1$ is eliminated. Let $\bigotimes' = \texttt{simulate}(\bigotimes, \psi)$ denote the competition simulation, where $\bigotimes'$ is the set of winner structures.

There are some discussions. We employ the competition simulation after the enumeration. Many bad topological structures can be eliminated according to the simulator. Only the winner topological structures will keep on optimizing their calculation operations. Furthermore, we can control the number of winners at a low level. Thus, the competition simulation can substantially reduce the search cost and improve the search efficiency.

14.2.3 Experience-Reused CNAS

The combination of CNAS and reused experience is the experience-reused CNAS algorithm shown in Algorithm 14.1. $\tilde{\mathcal{N}}$ denotes the best-so-far network architecture. At line 1, $\tilde{\mathcal{N}}$ is set as empty because the search has just begun. The loop from lines 2–33 is the block-based search. Line 3 is the topological structure enumeration. $\bigotimes =$

Algorithm 14.1 Experience-Reused CNAS

Input:
$\alpha, \beta, \mathcal{F}, N$: depth, width, operation space, block size
B, B': budget in each round and final optimization
$\boldsymbol{\phi}, \psi, \mathcal{D}_{\text{train}}$: operation predictor, competition simulator, training dataset
`enumerate`, `simulate`: enumeration, competition simulation sub-procedure
`initialize`, `compete`: initialization, competition sub-procedure

Procedure:

```
1: $\tilde{\mathcal{N}} = \emptyset$
2: for $t = 1$ to $N$ do
3:   $\bigotimes$ = simulate(enumerate($\alpha, \beta$), $\psi$), $\Omega = \emptyset$
4:   for each $\otimes \in \bigotimes$ do
5:     $\tilde{F}_{\otimes} = \boldsymbol{\phi}(\otimes)$
6:     $v = C_{\text{valid}}(\tilde{\mathcal{N}} + \otimes \triangleleft \tilde{F}_{\otimes}, \mathcal{D}_{\text{train}})$
7:     $\omega_{\otimes}$ = initialize($\tilde{F}_{\otimes}, v, \mathcal{F}_{\otimes}$)
8:     $\Omega = \Omega \cup \{\omega_{\otimes}\}$
9:   end for
10:  while $|\Omega| > 1$ do
11:    for each $\omega_{\otimes} \in \Omega$ do
12:      for $i = 1$ to $B$ do
13:        $F = \omega_{\otimes}$.sample($\mathcal{F}_{\otimes}$)
14:        $v = C_{\text{valid}}(\tilde{\mathcal{N}} + \otimes \triangleleft F, \mathcal{D}_{\text{train}})$
15:        $\omega_{\otimes} = \omega_{\otimes}$.update($F, v$)
16:        if $v > C_{\text{valid}}(\tilde{\mathcal{N}} + \otimes \triangleleft \tilde{F}_{\otimes}, \mathcal{D}_{\text{train}})$ then
17:          $\tilde{F}_{\otimes} = F$
18:        end if
19:      end for
20:    end for
21:    $\Omega$ = compete($\Omega$)
22:  end while
23:  $\omega_{\otimes} \leftarrow \Omega$
24:  for $i = 1$ to $B'$ do
25:    $F = \omega_{\otimes}$.sample($\mathcal{F}_{\otimes}$)
26:    $v = C_{\text{valid}}(\tilde{\mathcal{N}} + \otimes \triangleleft F, \mathcal{D}_{\text{train}})$
27:    $\omega_{\otimes} = \omega_{\otimes}$.update($F, v$)
28:    if $v > C_{\text{valid}}(\tilde{\mathcal{N}} + \otimes \triangleleft \tilde{F}_{\otimes}, \mathcal{D}_{\text{train}})$ then
29:      $\tilde{F}_{\otimes} = F$
30:    end if
31:  end for
32:  $\tilde{\mathcal{N}} = \tilde{\mathcal{N}} + \otimes \triangleleft \tilde{F}$
33: end for
34: return $\tilde{\mathcal{N}}$
```

$\{\otimes_1, \otimes_2, \ldots, \otimes_M\}$ denotes the set of possible structures. Line 3 is the competition simulation that aims to eliminate topological structures according to the experience. Lines 4–9 initialize the calculation operation optimization procedure, i.e., $\omega_{\otimes}$ for each topological structure $\otimes \in \bigotimes$. The initialization step employs the calculation operation prediction to get the operation configuration (line 5). Here, $\tilde{F}_{\otimes}$ denotes the best-so-far calculation operations for $\otimes$. Lines 10–22 are the competition mechanism. Each $\omega_{\otimes}$ will be pushed forward for B steps, and then it competes with each other until we get the winner (line 23). The calculation operations on the winner will be further optimized for B' steps. The final $\otimes \lhd \tilde{F}$ is the best network architecture in this block. Then, we combine it with the previous block architecture (line 32). After search procedures on all N blocks are finished, $\tilde{\mathcal{N}}$ is the winner network architecture and returned.

14.3 Empirical Study

We conduct our experiments on two style tasks. First, the benchmark image classification task is employed to investigate our method by comparing it with other state-of-the-art NAS methods. Then, we apply our method to automatically design the network architecture for the image denoising task. For each task, we first show the basic information about the dataset. Then, the implementation details are shown, which include the architecture space, evaluation settings, search settings, etc. Finally, we provide the result analysis for each dataset.

14.3.1 Image Classification Tasks

Implementation Details

For the CNAS algorithm, the architecture search space includes the topological space and the calculation operation space. In image classification tasks, we employ block-based search to reduce the size of the architecture space while still obtaining deeper network architectures for better performance. Specifically, we set a total of XX blocks for CIFAR-10... In each block, the depth and width of the micro-architecture are set as 5 and 2. We search the network architecture block by block. The search processes will repeat several times. In practice, the search efficiency is unsatisfactory in this way. To tackle this, we follow a popular approach, i.e., repeated blocks, which is widely used in NAS methods such as ENAS [11], DARTS [10], etc. The blocks are classified into two categories: normal block and reduction block. CNAS only searches two blocks sequentially, then stacks them repeatedly. Figure 14.4 illustrates the high-level structure of block-based search on image classification tasks. The calculation operation space gives the possible operations for each node in the topological structure. CNAS follows the operation setting of DARTS, which contains 6 operations: separable convolution with 3×3 and 5×5 kernels, dilated separable convolution with 3×3 and 5×5 kernels, average pooling with 3×3 kernel, and max pooling with 3×3 kernel. The skipping connect space is set as $\{\emptyset, 1, 2\}$. The operation space is set the same for both normal and reduction blocks, but the stride is set as 2 for reduction blocks. And the reduction blocks are only set at $\frac{1}{3}$ and $\frac{2}{3}$ of the depth of the whole network architecture. On the evaluation part, the criterion is accuracy on the validation data. Each network architecture that is sampled by CNAS has to be trained and validated. Due to the limited search total time, it is impossible to employ

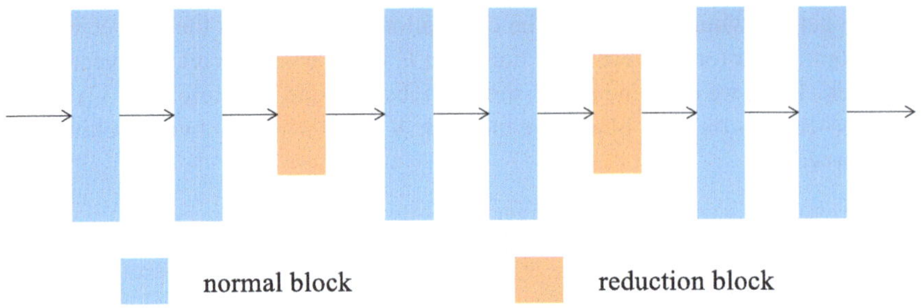

Fig. 14.4 Illustration of high-level structure for the block-based search on image classification tasks

the whole training data to evaluate an architecture. There are two ways to reduce the evaluation time cost: applying a part of the training data and applying few training epochs. We have tried some combinations of both of them, such as whole data with few epochs, a small part of data with large epochs, and a small part of data with few epochs. The empirical results show that the way of whole data with few epochs can obtain the rank performance of architectures. That is to say, evaluation from training architecture on the whole data but with few epochs may not obtain the final accuracy, but the architecture with good accuracy obtained by this way will perform well in the final testing. Thus, during the training process of the evaluation part, we employ whole data. The optimizer for training architectures is SGD with momentum 0.9 and weight decay 0.0003. The learning rate is set as 0.025. The batch size is 96. In the calculation operation optimization, we use SRacos (Chap. 6) as the search method, whose efficiency, stability, and scalability have been proved in many real AutoML applications. The total sample size is set as 200 for each experiment.

14.3.1.1 Result Analysis

We finish the experiments of CNAS on CIFAR-10 and CIFAR-100 on Tesla V100 GPU. We show the results on CIFAR-10 in Table 14.2. CNAS obtains 97.41% test accuracy, which outperforms the manual architecture DenseNet by more than 1% and outperforms the state-of-the-art NAS method DARTS by more than 0.2%. On the architecture size (parameter size), the size which CNAS obtains is much less than the size of the manual architecture (DenseNet). CNAS has a similar architecture size as DARTS's. From the above results, CNAS outperforms previous NAS methods on architecture quality (test accuracy and parameter size) and search efficiency (time cost). We show the results on CIFAR-100 in Table 14.3. DenseNet obtains 82.82% test accuracy on CIFAR-100. ENAS and DARTS fail to outperform DenseNet. CNAS obtains 83.03% test accuracy on CIFAR-100 by searching on it directly, which outperforms DenseNet.

Table 14.2 Comparisons of the architectures obtained by the state-of-the-art NAS approaches on CIFAR-10. † CNAS uses the evaluation policy with a partial dataset and many training epochs. ‡ CNAS uses the evaluation policy with the entire dataset and few training epochs

Architecture	Test Acc. (%)	Params (M)	Search cost (GPU days)	Method cate
ResNet + cutout	96.01	6.6	–	Manual
DenseNet + cutout	96.54	26.2	–	Manual
PNAS	96.59±0.09	3.2	225	SMBO
AmoebaNet + cutout	96.66±0.06	3.2	3150	Evolution
NASNet-A + cutout	97.35	3.3	1800	RL
ENAS-macro + cutout	96.13	38	0.32	RL
ENAS-micro + cutout	96.15	4.3	0.33	RL
NAONet	96.82	10.6	200	NAO
DARTS (first order) + cutout	97.00	3.3	1.50	Gradient-based
DARTS (second order) + cutout	97.24	3.3	4.00	Gradient-based
SNAS (mild constraint) + cutout	97.02	2.9	1.50	Gradient-based
SNAS (moderate constraint) + cutout	97.15	2.8	1.50	Gradient-based
SNAS (aggressive constraint) + cutout	96.90	2.3	1.50	Gradient-based
CNAS† + cutout	97.31	3.0	1.02	Competition
CNAS‡ + cutout	97.41	2.8	2.86	Competition

Table 14.3 Comparisons of the architectures evaluated on CIFAR-100. The column Search Dataset shows on which dataset the architecture is searched. If this column is CIFAR-10, it means the architecture is searched on CIFAR-10 and transferred to CIFAR-100

Architecture	Search dataset	Test Acc. (%)	Params (M)	Search cost (GPU days)	Method cate.
ResNet + cutout	–	78.04	6.6	–	Manual
DenseNet + cutout	-	82.82	26.2	–	Manual
ENAS-micro + cutout	CIFAR-10	81.28	4.3	0.32	RL
ENAS-micro + cutout	CIFAR-100	81.26	3.1	0.33	RL
DARTS + cutout	CIFAR-10	82.24	3.3	1.50	Gradient-based
DARTS + cutout	CIFAR-100	81.60	2.5	4.20	Gradient-based
CNAS + cutout	CIFAR-10	82.78	2.8	2.86	Competition
CNAS + cutout	CIFAR-100	83.03	3.2	2.08	Competition

14.3.2 Image Denoising Tasks

For image denoising tasks, we select the Waterloo and SIDD datasets. Waterloo has 3320 training images and 1423 testing images. SIDD has 113 training images and 48 testing images. We randomly add Gaussian noise to those images to construct the training datasets. We use a 50×50 window to clip images for Waterloo and SIDD. Then, we obtain 355200 training images on Waterloo and 739400 training images on SIDD.

Implementation Details

The image denoising task requires that the output image should have the same size as the input image. Thus, we have to constrain the macro architecture when searching. In this experiment, we select UNet as the macro architecture, which is shown in Fig. 14.5. The macro architecture has a symmetrical structure. The block that NAS searches should have the symmetry too, so the first block should have the same operations as the last block. For example, if the operations of the first block are $\{f_1, f_2, f_3\}$, then the operations of the last block are $\{f_3, f_2, f_1\}$. There are a total of 6 blocks. According to the symmetry, we should only search 3 blocks. CNAS uses block-based search to search block by block. For the topological structure, the depth of a block is 3, and the width is 1. For the operation space, blocks share the same operation space. But in each block, there are some differences: the operation space of the 1st block is 2×2, 3×3, and 5×5 convolutional operations with kernel $\{16, 32, 64\}$, the oper-

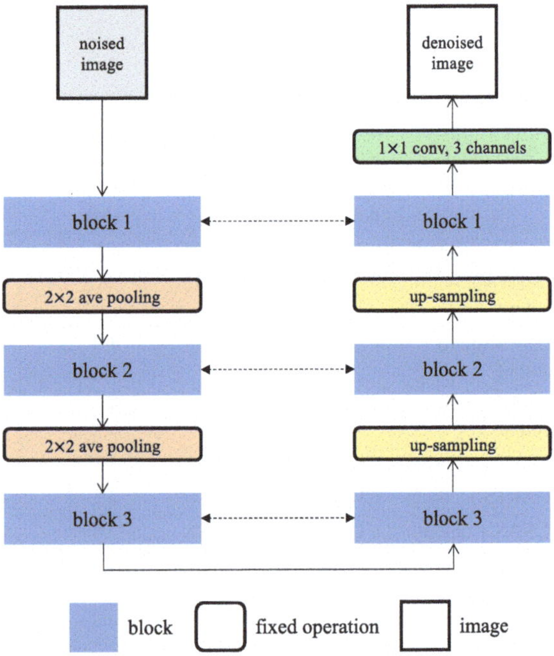

Fig. 14.5 Illustration of high-level structure for the block-based search on image denoising tasks

Table 14.4 Comparisons of the architectures evaluated on Waterloo and SIDD. The column Search Dataset shows on which dataset the architecture is searched

Dataset	Methods	PSNR (dB)	Params (M)	Search cost (GPU days)	Method cate.
Waterloo	CBDNet	34.57	4.13	–	Manual
	DnCNN	34.11	0.63	–	Manual
	CNAS	**34.68**	3.22	0.88	Competition
SIDD	CBDNet	38.11	4.13	–	Manual
	DnCNN	36.58	0.63	–	Manual
	CNAS	**38.15**	4.62	1.58	Competition

ation space of the 2nd block is 2×2, 3×3, and 5×5 convolutional operations with kernel $\{32, 64, 128\}$, and the operation space of the 3rd block is 2×2, 3×3, and 5×5 convolutional operations with kernel $\{64, 128, 256\}$. We select the Adam optimizer to train and test on Waterloo and SIDD. In the evaluation phase, we use $\frac{1}{7}$ of the dataset to train and $\frac{1}{10}$ of the dataset to validate. The learning rate of Adam is 0.0001. The batch size is 200. The training epoch is 20. In the testing phase, we use the whole dataset to train for 800 epochs. The learning rate of Adam is 0.00001. The evaluation criterion is the Peak Signal to Noise Ratio (PSNR) (Table 14.4).

14.3.2.1 Result Analysis

For the image denoising tasks, we select two manual architectures: CBDNet and DnCNN. The topological structure is pre-defined. Thus, we only test the operation optimization performance of CNAS. CNAS searches architectures based on the UNet macro architecture and obtains better PSNR than CBDNet on two datasets. On Waterloo, CNAS obtains an architecture that outperforms CBDNet on PSNR, and the parameter size is smaller than CBDNet's. On SIDD, CNAS outperforms CBDNet on PSNR, but the parameter size is a little larger than CBDNet's. The results demonstrate the effectiveness of CNAS.

14.4 Summary

This chapter presents a competition neural architecture search method (CNAS). The network architecture can be seen as a DAG that consists of a topological structure and calculation operations. CNAS considers the search on the topological structure and the calculation operations separately. For the topological structure, CNAS enumerates all possible structures under a limitation of depth and width. For each topological structure, CNAS considers the calculation operations as a black-box optimization problem and solves it by DFO methods, introduced in the previous chapters. Then, a competition mechanism is employed to combine both of them. To improve the search efficiency, this chapter applies a block-based search approach to constrain the search space. This chapter presents an experience reuse approach to search architectures by considering the manual experience and search experience of history. For operation optimization, the calculation operation prediction is proposed to predict high-quality operations for a topological structure. For topological enumeration, the competition simulation is proposed to select high-quality topological structures faster. In experiments on image classification and image denoising, the empirical results verify the effectiveness of CNAS.

References

1. Baker B, Gupta O, Naik N, Raskar R (2017) Designing neural network architectures using reinforcement learning. In: Proceedings of the 5th international conference on learning representations
2. Bengio Y (2000) Gradient-based optimization of hyperparameters. Neural Comput 12(8):1889–1900
3. Bergstra J, Bengio Y (2012) Random search for hyper-parameter optimization. J Mach Learn Res 13:281–305
4. Fogel DB (1994) An introduction to simulated evolutionary optimization. IEEE Trans Neural Netw 5(1):3–14
5. He K, Zhang X, Ren S, Sun J (2016) Deep residual learning for image recognition. In: Proceedings of the IEEE conference on computer vision and pattern recognition, pp 770–778
6. Howard AG, Zhu M, Chen B, Kalenichenko D, Wang W, Weyand T, Andreetto M, Adam H (2017) Mobilenets: Efficient convolutional neural networks for mobile vision applications. arXiv:1704.04861
7. Hu YQ, Qian H, Yu Y (2017) Sequential classification-based optimization for direct policy search. In: Proceedings of the 31st AAAI conference on artificial intelligence, San Francisco, CA, pp 2029–2035,
8. Hutter F, Hoos HH, Leyton-Brown K (2011) Sequential model-based optimization for general algorithm configuration. LION 5:507–523
9. Liu C, Zoph B, Neumann M, Shlens J, Hua W, Li LJ, Fei-Fei L, Yuille A, Huang J, Murphy K (2018) Progressive neural architecture search. In: Proceedings of the European conference on computer vision, pp 19–34
10. Liu H, Simonyan K, Yang Y (2019) DARTS: Differentiable architecture search. In: Proceedings of the 7th international conference on learning representations
11. Pham H, Guan M, Zoph B, Le Q, Dean J (2018) Efficient neural architecture search via parameter sharing. In: Proceedings of the 35th international conference on machine learning, pp 4092–4101
12. Real E, Aggarwal A, Huang Y, Le QV (2019) Regularized evolution for image classifier architecture search. In: Proceedings of the 33rd AAAI conference on artificial intelligence
13. Sandler M, Howard A, Zhu M, Zhmoginov A, Chen LC (2018) Mobilenetv2: inverted residuals and linear bottlenecks. In: Proceedings of the IEEE conference on computer vision and pattern recognition, pp 4510–4520
14. Shahriari B, Swersky K, Wang Z, Adams RP, Freitas ND (2015) Taking the human out of the loop: a review of Bayesian optimization. Proc IEEE 104(1):148–175
15. Szegedy C, Ioffe S, Vanhoucke V, Alemi A (2017) Inception-v4, inception-resnet and the impact of residual connections on learning. In: Proceedings of the 31st AAAI conference on artificial intelligence

16. Xie L, Yuille A (2017) Genetic CNN. In: Proceedings of the IEEE international conference on computer vision, pp 1379–1388
17. Xie S, Girshick R, Dollár P, Tu Z, He K (2017) Aggregated residual transformations for deep neural networks. In: Proceedings of the IEEE conference on computer vision and pattern recognition, pp 1492–1500
18. Yao X (1999) Evolving artificial neural networks. Proc IEEE 87(9):1423–1447
19. Zhong Z, Yan J, Wu W, Shao J, Liu CL (2018) Practical block-wise neural network architecture generation. In: Proceedings of the IEEE conference on computer vision and pattern recognition, pp 2423–2432
20. Zoph B, Le QV (2017) Neural architecture search with reinforcement learning. In: Proceedings of the 5th international conference on learning representations
21. Zoph B, Vasudevan V, Shlens J, Le QV (2018) Learning transferable architectures for scalable image recognition. In: Proceedings of the IEEE conference on computer vision and pattern recognition, pp 8697–8710

The manufacturer's authorised representative in the EU is Springer Nature Customer Service Centre GmbH, Europaplatz 3, 69115 Heidelberg, Germany. If you have any concerns regarding our products, please contact ProductSafety@springernature.com

Printed and bound by CPI Group (UK) Ltd, Croydon, CR0 4YY

01/10/2026

02225922-0002